Statistics
Problems and Solutions

Statistics
Problems and Solutions

E. E. Bassett
J. M. Bremner
I. T. Jolliffe
B. Jones
B. J. T. Morgan
P. M. North

Edward Arnold

First published in Great Britain 1986 by
Edward Arnold (Publishers) Ltd, 41 Bedford Square, London WC1B 3DQ

Edward Arnold (Australia) Pty Ltd, 80 Waverley Road, Caulfield East,
Victoria 3145, Australia

Edward Arnold, 3 East Read Street, Baltimore, Maryland 21202, U.S.A.

ISBN 0 7131 3568 9

Printed in Great Britain by J. W. Arrowsmith Ltd, Bristol

Preface

The idea for this book arose from a short course the authors have been giving at the University of Kent annually since 1974, to an audience consisting mainly of schoolteachers. The course has several aims, but one of the most important is to give the participants guided experience in solving problems in statistics. We have found that many people have a passing acquaintance with the various concepts encountered in a typical introductory statistics course for those with a basic knowledge of calculus: the sort of course which in Britain might be given to sixth-form mathematicians or to first year undergraduates. By contrast, people usually have little experience in applying the concepts in practical examples, and hence lack confidence.

These attributes of experience and confidence are vital to teachers and students alike, and since they can be gained most effectively by working through examples, we feel that a book of worked solutions to problems, containing many notes and comments, will be found to be of value.

Since this book is not designed as a standard textbook, we have in general not included straightforward expository material, even if a fair proportion of examination questions require 'bookwork' answers. By and large, such answers can be obtained most effectively through reading a good textbook, and we have framed most of the problems so as to omit these matters. Occasionally, however, inclusion of a 'bookwork' section of a problem has enabled us to widen horizons somewhat by discussing techniques slightly beyond most syllabuses, and we have felt it helpful to do this.

Since most readers of this book will have interests in solving problems set in public or professional examinations at A-level or first year undergraduate level, we have devised the problems after perusing many examination questions set at this level. The problems are all original, although we have in many (but not all) cases ensured that they are realistic in standard and style by keeping actual questions in mind while devising them; at the same time, we have of course taken care to respect the copyright in published examination questions. In this way we aim to give readers an insight into how professional statisticians view and deal with the sorts of problems set as questions in public examinations. Some relatively difficult problems (or parts of problems) have been marked with asterisks. Readers may prefer to postpone working through these parts until they have become familiar with the more straightforward material.

In one respect, though, that of length, the problems included in this book are often not representative of examinations questions. As we noted above, bookwork sections of questions have typically been excluded. But in general we have, in any case, preferred not to restrict ourselves to 'standard' length problems. In particular, when we have felt it desirable to explore the links between related statistical techniques we have extended a problem to well beyond acceptable examination length.

A quick glance through the pages of this book will reveal the authors' approach. When statistics is compared with, say, mechanics as an examination subject, we find that the algebraic manipulations are relatively straightforward, but the real difficulty comes from determining which techniques to apply to which problem. Because of this, straightforward solutions to problems are of limited value unless they contain a discussion of why the technique used is the most appropriate one. We have therefore deliberately elaborated solutions into more than bald 'model answers', and have in almost all cases provided extensive notes to the solutions, which we hope will add to the value of the solutions themselves. Some notes give alternative methods of solution, while others present solutions to related problems. We feel that it will often be from examining the notes that the reader will discover the relationships between the various statistical techniques which will give the confidence needed to tackle new problems. Some of the notes are more advanced in standard; as with the more difficult problems these have been marked with asterisks.

The most basic division in Statistics is that between probability and inference, and accordingly we use such a division in this book. One can describe the distinction between these two areas in a variety of ways. Perhaps the simplest is to note that problems of inference start with sample information and aim to make statements about the groups or populations from which the samples were randomly drawn. Thus, we have incomplete information about a population, and wish to infer what we can about that population.

A straightforward, if informal, way of starting this process is to construct a histogram from the sample. In doing this we aim to discover from what distribution the data might have come. Having completed such a preliminary analysis, and perhaps having performed a goodness-of-fit test, we may assume that a variable has, say, a normal distribution, but with the parameters of that distribution unknown. In such a case the inferences are then required only about those parameters. Other examples of inference are found in problems of regression and correlation in which a random sample gives partial information about the relationship between two variables. In inference, therefore, we have knowledge about part of a population, and we can regard inference as a form of generalisation.

Probability, by contrast, is the reverse process of arguing from general knowledge to special cases, a process of deduction. Since making deductions from axioms is the basic procedure of pure mathematics, probability is essentially a branch of mathematics, that branch dealing with problems with random or haphazard elements. So, typically, problems in probability theory describe the background, state where the haphazard (or random) features occur, and, generally, require one to calculate the probability of some event of interest.

The link between probability and statistical inference is just as clear-cut as is the distinction between them. Since inference attempts to make statements about entire populations on the basis of the partial information given by a sample, these statements cannot be both definitive and guaranteed as true. For example, if we test two fertilisers on a sample of crops, we cannot be certain that the one which appears to be the better will always give superior results; the most we can do is to couch our conclusion in terms of probability. Thus the statistician cannot do without results in probability theory in making inferences.

As with most examination questions, the problems devised for this book often do not fall neatly into single categories. It follows that, while the problems have been grouped according to the topic of primary interest, many problems have facets which take them out of their section, and sometimes indeed into a different chapter. Consequently, there are several occasions in early chapters in which reference is made to later material for further details, or for related material. The fact that statistical problems are so interdependent emphasises the unity of the subject; it is not just a collection of isolated techniques, but a comprehensive approach to the analysis of data.

A textbook is designed to be read from beginning to end, in sequence. By contrast, we expect readers to dip into this book; it is not intended to be the book which cannot be put down, but we hope it is the one which when picked up helps to resolve difficulties about how a technique is carried out and when it is appropriate. The only caution we would give a reader who does wish to dip into the book is that the introductions to individual chapters and sections should normally be read along with the problems, since notation and terminology are usually introduced there.

We have taken a common-sense approach to the matter of the accuracy to which calculations are carried out. Our aim is almost always to ensure that the result of an entire sequence of calculations is presented to a reasonable number of significant digits, and is accurate to the number of digits quoted. (There are one or two exceptions to this rule, for cases in which results are conventionally quoted to a fixed accuracy only.) Thus, for example, since tables of

Student's t-distribution on 10 degrees of freedom are generally given to three or four significant digits, it is pointless calculating a corresponding statistic to any greater accuracy, but it would be rather risky to be much less accurate in our calculations.

In most cases we have presented some intermediate values in a sequence of calculations. A difficulty in doing this is that if such a value is presented only to the number of significant digits in the initial or final values, the sequence of calculations may not appear to be correct. (As an unrealistically simple example, suppose we were evaluating $10/\sqrt{10}$ to 2 significant digits, without using the obvious method of cancelling! We would have to write

$$\frac{10}{\sqrt{10}} = \frac{10}{3{\cdot}2} = 3{\cdot}2,$$

but, of course, dividing 10 by 3·2 gives 3·125.)

Our practice in this book has been to present any intermediate steps to a reasonable accuracy, as a help to the reader wishing to follow the calculations through. But subsequent calculations are based not on the value presented but on a more accurate one; it is basically as if one was using a calculator which held the intermediate value to 8 significant digits, and one recorded a value only to a few significant digits as an aide-memoire.

A student completely unfamiliar with a subject does not start to study it by reading a book of worked examples. Accordingly, we have written for those who have read a first-level textbook such as *A Basic Course in Statistics*, by G. M. Clarke and D. A. Cooke (Edward Arnold, London, Second Edition, 1983), and will therefore be familiar with the terminology and most of the concepts of elementary probability and statistics. We do not set out to teach systematically matters like conditional probability, use of normal tables, histograms, t-tests and correlation (just to take a single example from each chapter). We assume that the reader has encountered these, and either knows what is involved or has a favourite textbook in which to look them up.

Finally, no book is ever the fruit of the effort and imagination of the authors alone. While we naturally accept full responsibility for any errors that may remain, we are grateful to all whose perceptive comments have helped to sharpen our own ideas about suitable ways to present and solve these problems. In particular, we warmly acknowledge the extensive questioning directed at us by the participants on our courses over the years, without whom we would not have embarked on the task of producing full solutions to all these problems.

Different authors have contributed different sections to this book, but we have shared in the production of all parts of the volume. Inevitably, some authors have contributed more than others, and in particular the task of overall editorial responsibility was mainly undertaken by one of us (E.E.B.).

The production of the book owes a great deal to the magnificent service offered by the Computing Laboratory at the University of Kent, and especially the Electronic Publishing Research Unit established by Mrs Heather Brown. The book was typed at a computer terminal, using the *troff* system, and the authors are most grateful for the provision of these facilities, without which the successive revisions of the material would have been very much more difficult.

Eryl Bassett
J. Michael Bremner
Ian Jolliffe
Byron Jones
Byron J. T. Morgan
Philip M. North

Canterbury,
June, 1986

Contents

1 Probability and Random Variables

As Professor Sir John Kingman remarked in a review lecture in 1984 on the 150th anniversary of the founding of the Royal Statistical Society,

'The theory of probability lies at the root of all statistical theory'.

The remark was challenged and led to an interesting discussion, but it is unquestionable that, at an elementary level, almost all statistical analyses depend for their validity on results in probability. (Exceptions are graphical methods – construction of histograms and so on; and indeed it is to a substantial extent the growth of less formal, graphically based, techniques in more advanced areas which made the above remark controversial.) It is natural, therefore, that methods based upon probability theory appear throughout this book.

From Chapter 3 onwards, probability will be just a means to an end, a tool to enable us to make valid statistical inferences. But in this chapter and the next we examine it more for its own sake. The first section of this chapter is devoted to a discussion of problems involving the manipulation of events and their corresponding probabilities, and the use of the basic laws and theorems of probability theory. Later, in Section 1B, we cover problems involving random variables and their distributions, paving the way for the coverage of well-known probability distributions in Chapter 2.

1A Probability

The theory of probability has, as its central feature, the concept of a repeatable random experiment; that is, an experiment the outcome of which is uncertain. Obvious examples are simple games using dice or playing cards. The set of all possible outcomes of an experiment is known as its *sample space* (another term which is sometimes used is *possibility space*), and subsets of this are called *events*. In many problems, the aim will be to calculate the probability associated with some event of interest.

Very often the sample space will have an easily recognised structure, and one can then exploit this structure mathematically to calculate any required probabilities. The most common structures of this sort occur when a sample space is a set of integers, or a set of real numbers, and these are examined in Section 1B. In the present section we deal with problems in which we have no such advantage. The tools used in this section, therefore, are the general ones of probability theory, applicable in all cases. In particular we will need to use the addition and multiplication laws, and the law of total probability and Bayes' Theorem.

The problems set in this section generally require a numerical answer. For a calculation to be possible one must be able to assign a probability numerically to each of the outcomes of the experiment concerned. Some problems may relate to situations possessing sufficient symmetry for the assumption of 'equally likely outcomes' to be appropriate; calculation of probabilities then reduces simply to a question of counting outcomes. For other problems such an assumption may not be appropriate, and we must be given, directly or indirectly, values of the probabilities of various events. The first six problems in this section are of the former type, while the remaining five are of the latter.

Probability problems are often found difficult. Whereas a typical problem in another area will fall into one of a range of standard categories, so that a standard method of solution will be appropriate, this is less often true of probability problems. Indeed, it is frequently the case that there are two or more possible approaches to the solution of a problem, and that none of these is clearly 'best'; the choice of an approach is often largely a matter of personal taste. We have, therefore, presented more than one solution to several of the problems in this section. This is, in part, to demonstrate the range of different approaches that can be used. Another reason for presenting multiple solutions is that they illustrate a method of checking solutions (a familiar experience of those struggling with probability problems is that of reaching what turns out to be the wrong answer through an apparently flawless argument). If two solutions, using different methods, produce the same answer there is some hope that it may be correct! The luckless examination candidate is, unfortunately, unlikely to have time for this.

We use a natural and conventional notation. The probability of an event will be denoted by Pr(*xxxx*), where '*xxxx*' represents a verbal statement defining the event. A notation as simple as this suffices for some cases, but when the description of events is complex we resort to notation for the events themselves. Thus we may specify that an event will be denoted by a symbol – a capital letter, perhaps with a subscript, e.g. E_n – and we will then, of course, write $\Pr(E_n)$ for its probability. Finally, we denote the conditional probability of A given B by the standard notation $\Pr(A \mid B)$.

1A.1 The wellington boots

Three young children are attempting to retrieve their wellington boots from a cupboard. Each selects two boots at random; the cupboard contains no other boots.

(a) Find the probability that each obtains the correct pair of boots.

(b) Find the probability that each obtains a pair of boots (not necessarily the correct pair).

(c) Find the probability that at least one obtains the correct pair of boots.

(d) Now suppose that two pairs of boots are red, and one pair is yellow, and that the children select boots of the correct colour and put them on (without distinguishing between left and right). Find the probability that all three children end up wearing their own boots correctly.

Solution

For the first three parts of the question we shall adopt a sample space consisting of all possible allocations of two boots to each of the three children. There are $\frac{6!}{2!\,2!\,2!} = 90$ such allocations, each with probability $\frac{1}{90}$.

(a) Since only one allocation is correct, the probability that each child obtains the correct pair of boots is $\frac{1}{90}$.

(b) If we denote the three children by A, B, and C and let, for example, *CAB* stand for the outcome in which A obtains the boots belonging to C, B obtains those belonging to A, and C obtains those belonging to B, there are six outcomes in which each child obtains a pair, viz. *ABC*, *ACB*, *BAC*, *BCA*, *CAB*, *CBA*. Thus the probability of this event is $\frac{6}{90} = \frac{1}{15}$.

(c) We consider first allocations such that A obtains the correct pair of boots, while B and C do not. To count these, we note that, since A is to have the correct boots, we need only consider allocations of boots to B and C. There are $\binom{4}{2} = 6$ of these, of which all but one result in B and C obtaining incorrect pairs of boots. Thus there are 5 allocations with the property that only A obtains the correct boots; similarly there are 5 allocations such that only B obtains the correct boots, and 5 such that only C does. Adding to these the single allocation such that they all obtain the correct boots, we find that there are 16 allocations which result in at least one child obtaining the correct boots, so that the required probability is $\frac{16}{90} = \frac{8}{45}$.

(d) We may consider the red and yellow boots separately. The probability that the child with the yellow boots puts them on correctly is $\frac{1}{2}$. For the red boots, we consider a sample space consisting of all 4! possible allocations of the boots to the feet of the children concerned (in this part of the question the order is important). Only one such allocation is correct, so that the probability that the red-booted children are correctly shod is $\frac{1}{4!} = \frac{1}{24}$. Finally, assuming independence between the allocation of red and yellow boots, we obtain the probability that all three children are correctly shod as $\frac{1}{2} \times \frac{1}{24} = \frac{1}{48}$.

Notes

(1) The words 'at random' are frequently encountered in questions such as this. Strictly speaking, they imply only that the outcome cannot be predicted in advance. The interpretation we have chosen here is much stronger, implying, for parts (a) and (b), that each of the 90 possible outcomes has the same chance of occurring. Such an assumption of 'equally likely outcomes' is in line with everyday usage in statements such as 'Winning Premium Bond numbers are chosen at random.' In solving probability problems, it is the natural one in the absence of any further information, and we shall frequently make it in our solutions to the problems which follow.

(2) It is possible to solve parts (a)-(c) on the basis of a sample space in which all 6! possible *ordered* allocations of the boots to the children are equally likely. Then, for example, there are 8 ($= 2^3$) outcomes in which all three children obtain the correct boots, since each child can obtain his (or her) boots in two possible orders.

(3) Some readers may prefer to tackle this problem by dividing each part into stages and using the multiplication laws of probability. We now present a solution along these lines.

(a) We may imagine A having first choice of boots, followed by B and finally C. If we adopt a notation in which, for example, A_1 denotes the event that the first boot chosen by A belongs to A, we require the probability of the event $A_1 \cap A_2 \cap B_1 \cap B_2 \cap C_1 \cap C_2$. This probability may be expressed as

$$\Pr(A_1)\Pr(A_2|A_1)\Pr(B_1|A_1 \cap A_2)\Pr(B_2|A_1 \cap A_2 \cap B_1)$$

(we have omitted two further conditional probabilities in order to shorten the above expression: they relate to C's 'choice', and are both equal to 1 since C has no choice at all). Now $\Pr(A_1) = \frac{2}{6}$, since A has initially 6 boots to choose from, of which 2 are correct; and $\Pr(A_2|A_1) = \frac{1}{5}$, since at this stage A has 5 boots to choose from, of which 1 is correct. Similarly, $\Pr(B_1|A_1 \cap A_2) = \frac{2}{4}$, and $\Pr(B_2|A_1 \cap A_2 \cap B_1) = \frac{1}{3}$. Multiplying these probabilities, we obtain the solution $\frac{1}{90}$, as before.

(b) If we let M_{A} stand for the event that A obtains a matching pair of boots, and so on, we require

$$\Pr(M_{\mathrm{A}} \cap M_{\mathrm{B}} \cap M_{\mathrm{C}}) = \Pr(M_{\mathrm{A}})\Pr(M_{\mathrm{B}}|M_{\mathrm{A}})$$

(as above, the third component $\Pr(M_{\mathrm{C}}|M_{\mathrm{A}} \cap M_{\mathrm{B}})$ is equal to 1, and has therefore been omitted). Now $\Pr(M_{\mathrm{A}})$ is just the probability that the second boot selected by A matches the first, i.e. $\frac{1}{5}$; similarly $\Pr(M_{\mathrm{B}}|M_{\mathrm{A}}) = \frac{1}{3}$. Multiplying these probabilities, we obtain the solution $\frac{1}{15}$.

(c) We now let A denote the event that A obtains the correct pair of boots, and define B and C similarly. We then require $\Pr(A \cup B \cup C)$. Using the general addition law for three events, this probability may be expanded as

$$\Pr(A) + \Pr(B) + \Pr(C) - \Pr(A \cap B) - \Pr(B \cap C) - \Pr(A \cap C) + \Pr(A \cap B \cap C).$$

Now, using the notation used in solving part (a), since $A = A_1 \cap A_2$, we obtain

$$\Pr(A) = \Pr(A_1)\Pr(A_2 | A_1) = \tfrac{2}{6} \times \tfrac{1}{5} = \tfrac{1}{15}.$$

Similarly, $\Pr(B)$ and $\Pr(C)$ have the same value.

The four remaining terms that we require in order to obtain $\Pr(A \cup B \cup C)$ are all the same, and equal to $\Pr(A \cap B \cap C)$ (since, for example, if A and B obtain the correct boots, so also must C). This probability has already been shown, in part (a), to be $\frac{1}{90}$. Thus

$$\Pr(A \cup B \cup C) = 3 \times \tfrac{1}{15} - 3 \times \tfrac{1}{90} + \tfrac{1}{90} = \tfrac{8}{45}.$$

(d) We shall use a notation in which, for example, L_A will stand for the event that the boot worn by A on his (or her) left foot is the correct one. We may, without loss of generality, take it that A is the owner of the yellow boots. Dealing first with these, we obtain

$$\Pr(L_A \cap R_A) = \Pr(L_A)\Pr(R_A | L_A) = \tfrac{1}{2} \times \tfrac{1}{1} = \tfrac{1}{2}.$$

Turning to the red boots, we obtain

$$\Pr(L_B \cap R_B \cap L_C \cap R_C) = \Pr(L_B) \times \Pr(R_B | L_B) \times \Pr(L_C | L_B \cap R_B) \times \Pr(R_C | L_B \cap R_B \cap L_C)$$
$$= \tfrac{1}{4} \times \tfrac{1}{3} \times \tfrac{1}{2} \times \tfrac{1}{1} = \tfrac{1}{24}.$$

Multiplying these two probabilities, we obtain the solution $\frac{1}{48}$, as before.

1A.2 Dealing four cards

Four cards are dealt from a standard pack of 52 cards. Find

(i) the probability that all four are spades;

(ii) the probability that two or fewer are spades;

(iii) the probability that all four are spades, given that the first two are spades;

(iv) the probability that spades and hearts alternate.

Solution

We shall solve this problem using three different approaches. Of these, Method 1 is probably the most straightforward and appealing.

Method 1

Let S_i denote the event that the ith card dealt is a spade, and H_i the event that it is a heart. The required probabilities are then obtained as follows:

(i) $$\Pr(S_1 \cap S_2 \cap S_3 \cap S_4) = \Pr(S_1)\Pr(S_2 | S_1)\Pr(S_3 | S_1 \cap S_2)\Pr(S_4 | S_1 \cap S_2 \cap S_3)$$
$$= \frac{13}{52} \times \frac{12}{51} \times \frac{11}{50} \times \frac{10}{49}$$
$$= \frac{11}{4165}.$$

(ii) Rather than sum the probabilities of the three events 'no spades', 'one spade', and 'two spades', it will be simpler to adopt the standard device of obtaining the probability of the event complementary to that specified in the problem, and then subtracting from 1. Adopting this approach, we have to calculate the probabilities of only two events; there is, indeed, a further reduction in effort, since we have already found one of these probabilities in the solution to part (i). Thus

$$\Pr(\text{two or fewer are spades}) = 1 - \Pr(\text{three are spades}) - \Pr(\text{all four are spades}),$$

and, since the second of the probabilities is known, we need only the following:

$$\Pr(\text{three are spades}) = \Pr(S_1 \cap S_2 \cap S_3 \cap \bar{S}_4) + \Pr(S_1 \cap S_2 \cap \bar{S}_3 \cap S_4) + \Pr(S_1 \cap \bar{S}_2 \cap S_3 \cap S_4) + \Pr(\bar{S}_1 \cap S_2 \cap S_3 \cap S_4).$$

The first of these four probabilities is

$$\Pr(S_1)\Pr(S_2 | S_1)\Pr(S_3 | S_1 \cap S_2)\Pr(\bar{S}_4 | S_1 \cap S_2 \cap S_3) = \frac{13}{52} \times \frac{12}{51} \times \frac{11}{50} \times \frac{39}{49}.$$

Similarly, the remaining three probabilities are

$$\frac{13}{52} \times \frac{12}{51} \times \frac{39}{50} \times \frac{11}{49}, \quad \frac{13}{52} \times \frac{39}{51} \times \frac{12}{50} \times \frac{11}{49}, \quad \frac{39}{52} \times \frac{13}{51} \times \frac{12}{50} \times \frac{11}{49}$$

respectively. Combining these results with the solution to part (i), we obtain

$$\Pr(\text{two or fewer are spades}) = 1 - \frac{(13 \times 12 \times 11 \times 10) + (4 \times 13 \times 12 \times 11 \times 39)}{52 \times 51 \times 50 \times 49} = \frac{19\,912}{20\,825}.$$

(iii) $$\Pr(S_1 \cap S_2 \cap S_3 \cap S_4 | S_1 \cap S_2) = \Pr(S_3 | S_1 \cap S_2)\Pr(S_4 | S_1 \cap S_2 \cap S_3) = \frac{11}{50} \times \frac{10}{49} = \frac{11}{245}.$$

(iv) The required probability is composed of two parts, $\Pr(S_1 \cap H_2 \cap S_3 \cap H_4)$ and $\Pr(H_1 \cap S_2 \cap H_3 \cap S_4)$. The first of these probabilities is

$$\Pr(S_1)\Pr(H_2 | S_1)\Pr(S_3 | S_1 \cap H_2)\Pr(H_4 | S_1 \cap H_2 \cap S_3) = \frac{13}{52} \times \frac{13}{51} \times \frac{12}{50} \times \frac{12}{49},$$

and the second is found, similarly, to have the same value. The required probability is thus

$$\frac{2 \times 13^2 \times 12^2}{52 \times 51 \times 50 \times 49} = \frac{156}{20\,825}.$$

Method 2

In this method we base our solutions on the enumeration of appropriate outcomes in a sample space consisting of all possible hands of four cards, taking the order in which the cards are dealt into account. This sample space consists of $52 \times 51 \times 50 \times 49$ outcomes – all equally likely.

(i) The number of different hands consisting of four spades is $13 \times 12 \times 11 \times 10$. Hence the probability that a hand consists entirely of spades is $\frac{13 \times 12 \times 11 \times 10}{52 \times 51 \times 50 \times 49}$, as before.

(ii) As in Method 1, it is simpler to count the number of hands with three or four spades and subtract. The number of different hands consisting of three spades followed by one card which is not a spade is $13 \times 12 \times 11 \times 39$; considering the three other positions in which the card which is not a spade can appear, we see that the number of relevant hands is the same for each. Hence the total number of hands containing three or four spades is $(13 \times 12 \times 11 \times 10) + (4 \times 13 \times 12 \times 11 \times 39)$, from which we can deduce the same answer as that obtained by Method 1.

(iii) In order to obtain the required conditional probability, we must find the probability that the first two cards are spades. Since the number of hands with this property is $13 \times 12 \times 50 \times 49$, it follows that the corresponding probability is $\frac{13 \times 12}{52 \times 51}$. Dividing the probability obtained in answer to (i) by this quantity, we obtain the required result.

(iv) The number of possible hands in which spades and hearts alternate, with a spade in the first position, is $13\times13\times12\times12$; the number of hands in which spades and hearts alternate, with a heart in the first position, is the same. We thus obtain the same probability as was obtained by Method 1.

Method 3

In this method (which can be used only for parts (i) and (ii) of the problem) we again consider a sample space of equally likely four-card hands, but do not take the order in which the cards are dealt into account. There are thus $\binom{52}{4}$ outcomes in the sample space.

(i) There are $\binom{13}{4}$ possible hands consisting of four spades. The probability of such a hand is thus

$$\frac{\binom{13}{4}}{\binom{52}{4}},$$

which, after appropriate cancellations, reduces to the solution already obtained.

(ii) To count the number of hands containing three spades and one card of another suit, we note that there are 39 possibilities for the latter card, and that each of these may be combined with $\binom{13}{3}$ possible combinations of three spades. The total number of hands is thus $39\times\binom{13}{3}$ and, dividing by $\binom{52}{4}$ and cancelling, we obtain the same answer as before.

Notes

(1) In Method 1, the introduction of an appropriate notation for the events of interest makes it possible to write out a fairly concise, yet clear, solution. While one might prefer a solution where the events are described verbally, this becomes somewhat lengthy. Consider, for example, part (i):

Pr(all four cards are spades)

= Pr(first card is a spade)

× Pr(second card is a spade | first card is a spade)

× Pr(third card is a spade | first two cards are spades)

× Pr(fourth card is a spade | first three cards are spades)

= . . . etc.

On the other hand, writing down $\Pr(SSSS)$, without explaining what this means, is not enough; and the inadequacy of such a notation becomes more evident when we start writing out a solution, involving meaningless terms like $\Pr(S\,|\,SS)$. (See Note 5 to Problem 1A.10 and Note 1 to Problem 1B.3 for a discussion of similar points.)

(2) In solving part (ii) by Method 1 it is tempting, in finding the probability of three spades, to obtain $\Pr(S_1 \cap S_2 \cap S_3 \cap \bar{S}_4)$ and then multiply by four, on the grounds that there are four possible positions for the card which is not a spade. What is more, this argument works! However, although the argument is valid for similar questions involving the rolling of dice, or the random selection of cards *with replacement* (and is fundamental to the derivation of the binomial distribution), it is not valid in situations where selection is performed *without replacement.* In the present case note that, although the four components which are summed to obtain the probability of three spades are equal, the probabilities which are multiplied to obtain them are not. A slightly more complicated argument, such as that we have given, is therefore necessary. (See Note 2 to Problem 1B.3 for a similar point, and for mention of a problem for

which the simpler argument does *not* give the correct answer.)

(3) When we are dealing with a sample space in which all outcomes are equally likely, we may use the fact that

$$\Pr(A \mid B) = \frac{n(A \cap B)}{n(B)},$$

where n denotes the number of outcomes in the indicated event. Thus, in solving part (iii) by Method 2, we have

$$n(S_1 \cap S_2 \cap S_3 \cap S_4) = 13 \times 12 \times 11 \times 10,$$

and

$$n(S_1 \cap S_2) = 13 \times 12 \times 50 \times 49,$$

leading to the same result as before.

* (4) As a further variant on the Method 2 solution to part (iii), we may consider a reduced sample space in which, given that the first two cards dealt are spades, there are 50×49 possibilities for the next two cards dealt – all of them equally likely. Of these, 11×10 consist of two spades, so that the required conditional probability is $\frac{11 \times 10}{50 \times 49}$, as before.

This argument is rather less straightforward than it may seem. We are, in fact, conditioning on the first two cards being a *particular* pair of spades (the three and the queen, say). Since the probability obtained does not depend on which pair we are conditioning on, it is also the probability conditional on the first two cards being *any* pair of spades.

* (5) While the approach of Method 1 is probably the simplest for this problem, that of Method 3 can be more easily extended to cover similar problems where larger numbers are involved. For example, the probability that, in a hand consisting of ten cards, six are spades is, by an argument similar to that used for part (ii),

$$\frac{\binom{13}{6}\binom{39}{4}}{\binom{52}{10}}.$$

We are dealing here with the *hypergeometric* distribution, which is discussed in greater detail in Problem 2A.11.

1A.3 The school assembly

At the morning assembly, five schoolchildren – Alan, Barbara, Clare, Daniel and Edward – sit down in a row along with five other children (whose names need not concern us here). If the children arrange themselves at random, find the probabilities of the following events.

(a) Alan and Barbara sit together.

(b) Clare, Daniel and Edward sit together.

(c) Clare, Daniel and Edward sit together but Alan and Barbara sit apart.

(d) Daniel sits between Clare and Edward (but not necessarily adjacent to either of them).

Solution

There are at least two ways of solving this problem. One is based on enumerating the seating arrangements corresponding to the events of interest, there being 10! equally likely arrangements in all. A second approach involves the multiplication of appropriate probabilities and conditional probabilities. We shall solve all four parts of the problem using the first approach (referred to as Method 1 below), and parts (a), (b), and (d) using the second (Method 2).

Method 1

(a) To count the number of possible arrangements here, we consider permutations of nine objects, one of which consists of A(lan) and B(arbara) sitting together, the other eight being the remaining children. The number of possible permutations of nine objects is 9! so that, after allowing for the fact that A may be sitting to the left, or to the right, of B, the total number of arrangements in which A and B sit together is $2 \times 9!$. The required probability is thus

$$\frac{2 \times 9!}{10!} = \frac{1}{5}.$$

(b) As in the solution to part (a), we consider permutations of eight objects: C(lare), D(aniel) and E(dward) sitting together, and the remaining seven children. Since there are 3! possible internal arrangements of the group of three children, and 8! ways of arranging the eight objects in order, the required probability is

$$\frac{3! \times 8!}{10!} = \frac{1}{15}.$$

(c) Direct enumeration is difficult here. We note, however, that we already know the probability that C, D and E sit together so that, if we can find the probability that they sit together *and* A and B sit together, we may obtain the required probability by subtraction. Formally,

$$\Pr(\overline{AB} \cap CDE) = \Pr(CDE) - \Pr(AB \cap CDE).$$

(Here, with a slight misuse of notation, we use AB to denote the event that A and B sit together and CDE the event that C, D and E sit together.) We find $\Pr(AB \cap CDE)$ by considering the 7! possible permutations of seven objects, one consisting of A and B, one consisting of C, D and E, and the other five being the remaining children. Taking into account the 2 possible orders for A and B and the 3! possible orders for C, D and E, we obtain

$$\Pr(AB \cap CDE) = \frac{2 \times 3! \times 7!}{10!} = \frac{1}{60}.$$

The solution to this part of the problem is therefore

$$\frac{1}{15} - \frac{1}{60} = \frac{1}{20}.$$

(d) Consider first the seven other children. They can be arranged in $10 \times 9 \times 8 \times \ldots \times 4$ ($= 10!/3!$) ways. For each such arrangement, there are 2 ways in of arranging C, D and E in the remaining seats with D sitting between C and E. Thus the solution to this part is

$$\frac{2 \times 10!/3!}{10!} = \frac{1}{3}.$$

Method 2

(a) If we consider the position of A, there are two different situations: when A is in an end position, and when A is in an internal position. The required probability is thus

$$\begin{aligned} &\Pr(\text{A in end position})\Pr(\text{B next to A} \mid \text{A in end position}) \\ &\quad + \Pr(\text{A in internal position})\Pr(\text{B next to A} \mid \text{A in internal position}) \\ &= \left(\frac{2}{10} \times \frac{1}{9}\right) + \left(\frac{8}{10} \times \frac{2}{9}\right) \\ &= \frac{1}{5}. \end{aligned}$$

(The conditional probabilities are obtained by regarding the nine possible positions for B, given that of A, as equally likely.)

(b) Let us consider first the event that C, D and E sit together in that order. For this to be true, C must be in one of the positions 1, 2, . . . , 8. If C is in position i $(1 \le i \le 8)$, the conditional probability that C, D and E sit together in that order is

$$\Pr(\text{D in position } i+1 \mid \text{C in position } i)\times$$
$$\Pr(\text{E in position } i+2 \mid \text{C in position } i,\ \text{D in position } i+1)$$
$$= \frac{1}{9}\times\frac{1}{8}.$$

Thus

$$\Pr(\text{C, D and E sit together in that order})$$
$$= \Pr(\text{C in one of positions } 1, 2, \ldots, 8)\times\frac{1}{9}\times\frac{1}{8}$$
$$= \frac{8}{10}\times\frac{1}{9}\times\frac{1}{8}$$
$$= \frac{1}{90}.$$

Finally, since there are 6 (i.e. 3!) possible orders for C, D and E to be considered, the required probability is $\frac{6}{90} = \frac{1}{15}$.

(c) A solution to this part using Method 2, although possible, is rather complicated, and is not provided here.

(d) We solve this part by first working conditionally on D's position, which must be one of 2, 3, . . . , 9; each of these has probability $\frac{1}{10}$. If D sits in position i, the conditional probability that C sits to the left of him and E sits to the right of him is

$$\frac{i-1}{9}\times\frac{10-i}{8};$$

the conditional probability that C sits to the right of him and E sits to the left has the same value. Thus the probability we require is

$$\sum_{i=2}^{9}\Pr(\text{D in position } i)\times\Pr(\text{C and E on opposite sides of D} \mid \text{D in position } i)$$
$$= \sum_{i=2}^{9}\frac{1}{10}\times\frac{2(i-1)(10-i)}{9\times 8}$$
$$= \frac{2(1\times 8\ +\ 2\times 7\ +\ \ldots\ +\ 8\times 1)}{10\times 9\times 8}$$
$$= \frac{240}{720} = \frac{1}{6}.$$

Notes

(1) We remarked in Note 5 to Problem 1A.2 that, while an approach based on splitting the experiment into stages (cf. Method 2 for this problem, and Method 1 for Problem 1A.2) may suggest itself as the most natural approach for many problems, it may turn out that some problems require us to enumerate outcomes in the sample space of the whole experiment. Although the latter approach may not appeal, because of the complicated combinatorial arguments that it can involve, it can sometimes prove (as in this problem) more powerful than the former.

(2) A simple solution to (d) follows from the observation that there are six possible orders in which C, D and E can sit, these being equally likely. Since two of these orders are such that D sits between C and E, the required probability is $\frac{2}{6} = \frac{1}{3}$.

1A.4 The second card

Two cards are dealt from a pack. Find the probability that the second card dealt is a heart.

Solution

The answer is clearly $\frac{1}{4}$, since all four suits must be equally likely. If this argument appears treacherous in its simplicity, the argument given in the following paragraph may be preferred.

If we let H_i $(i = 1, 2)$ denote the event that the ith card dealt is a heart then, using the law of total probability,

$$\begin{aligned}\Pr(H_2) &= \Pr(H_1)\Pr(H_2 | H_1) + \Pr(\bar{H}_1)\Pr(H_2 | \bar{H}_1) \\ &= \left(\frac{13}{52} \times \frac{12}{51}\right) + \left(\frac{39}{52} \times \frac{13}{51}\right) \\ &= \frac{1}{4}.\end{aligned}$$

Notes

(1) Newcomers to the study of probability theory frequently find this question confusing. A common answer when the question is put to a class is 'It depends on what the first card is' – but what is asked for is not a *conditional* probability, but the *unconditional* probability that the second card is a heart. The conditional probabilities appear in the second of the given solutions, where the unconditional probability is obtained as their weighted average.

(2) Even students who are clear about what is being asked for are reluctant to trust the first method of solution, preferring the second approach. This approach will not, however, extend easily to similar problems involving many cards. For example, suppose we require the probability that, when ten cards are dealt, the last one is a heart. While it is again clear that this is $\frac{1}{4}$, the sceptical will find that the previous approach has become rather cumbersome, involving the enumeration of 2^9 possibilities. They may, however, make use of an argument based on considering all of the $52 \times 51 \times \ldots \times 43$ possible hands of 10 cards (taking the order of dealing into account) to be equally likely. To count the number of such hands in which the final card is a heart we note that, for each of the 13 possibilities for the final card, there are $51 \times 50 \times \ldots \times 43$ different possibilities for the preceding nine. Thus the probability that the last card is a heart is

$$\frac{13 \times 51 \times 50 \times \ldots \times 43}{52 \times 51 \times 50 \times \ldots \times 43} = \frac{1}{4}.$$

* 1A.5 A birthday coincidence

The two authors of a letter to *The Times*, published in September 1984, wrote as follows:

> 'Whilst we were travelling by train back from work with two other friends, we happened to discover that the four of us had birthdays on three successive days, two being on the same day. Using our limited arithmetic and electronic calculators we have worked out the odds of this rare occurrence as being approximately 1,350,000 to 1.'

Verify the correctness of the stated odds.

Solution

As with several of the problems in this section, we shall present more than one solution, illustrating different possible approaches. In presenting each solution we shall refer to the four passengers as P_1, P_2, P_3 and P_4.

Method 1

We shall consider the four passengers in order, and introduce the following notation, relating to P_i. The event that this passenger has the *same* birthday as one of those previously considered will be denoted by S_i; A_i will denote the occurrence of a birthday *adjacent* to, i.e. one day before or after, one of the birthdays of those previously considered; C_i will denote the occurrence of a birthday *close* to, i.e. two days before or after, that of P_1, while we will let B_i stand for the event that P_i's birthday falls *between* that of P_1 and the 'close' birthday of another passenger. The event which aroused the interest of the letter-writers then occurs if the events relating to P_2, P_3 and P_4 are either

(i) two As and an S, in any order, or

(ii) a C, a B and an S, with the C preceding the B but the S in any position.

We shall consider these two possibilities in turn.

To obtain the probability of (i), we sum the probabilities of the three events $S_2 \cap A_3 \cap A_4$, $A_2 \cap S_3 \cap A_4$ and $A_2 \cap A_3 \cap S_4$. The first of these three events has probability

$$\Pr(S_2)\Pr(A_3 | S_2)\Pr(A_4 | S_2 \cap A_3) = \frac{1}{365} \times \frac{2}{365} \times \frac{2}{365} = \frac{4}{365^3}.$$

Similarly, the second and third events have probabilities

$$\Pr(A_2)\Pr(S_3 | A_2)\Pr(A_4 | A_2 \cap S_3) = \frac{2}{365} \times \frac{2}{365} \times \frac{2}{365} = \frac{8}{365^3}$$

and

$$\Pr(A_2)\Pr(A_3 | A_2)\Pr(S_4 | A_2 \cap A_3) = \frac{2}{365} \times \frac{2}{365} \times \frac{3}{365} = \frac{12}{365^3}$$

respectively. Totalling these, we find that the probability corresponding to (i) is $\frac{24}{365^3}$.

Turning now to (ii), we obtain the required probability as

$$\Pr(S_2 \cap C_3 \cap B_4) + \Pr(C_2 \cap S_3 \cap B_4) + \Pr(C_2 \cap B_3 \cap S_4)$$

$$= \left(\frac{1}{365} \times \frac{2}{365} \times \frac{1}{365}\right) + \left(\frac{2}{365} \times \frac{2}{365} \times \frac{1}{365}\right) + \left(\frac{2}{365} \times \frac{1}{365} \times \frac{3}{365}\right)$$

$$= \frac{12}{365^3}.$$

Finally, we add the probabilities corresponding to (i) and (ii) to obtain the result

$$\frac{36}{365^3} \approx \frac{1}{1\,350\,753},$$

giving odds against the observed coincidence of about 1 350 752 to 1.

Method 2

For the event of interest to occur it must be the case that two of the passengers have the same birthday, while the remaining passengers have birthdays which, along with the common birthday of the first two, fall on three consecutive dates. It will be sufficient to consider the case when P_1 and P_2 have the same birthday. When we have found the corresponding probability, symmetry considerations will enable us to obtain the solution to the problem by multiplying by 6 (since this is the number of distinct pairs amongst the four passengers).

Forgetting, for the moment, about P_2, the probability that P_1, P_3 and P_4 have birthdays on consecutive dates is obtained by adding the probabilities of two events. First, P_3's birthday may be one day before or after that of P_1, and P_4's birthday must then occur immediately before or after the pair of dates occupied by the birthdays of P_1 and P_3: the associated probability is $\frac{2}{365}\frac{2}{365} = \frac{4}{365^2}$. Second, P_3's birthday may be two days before or after that of P_1, in which case P_4's birthday must occur between the two dates: the probability of this is $\frac{2}{365}\frac{1}{365} = \frac{2}{365^2}$. Thus the probability that P_1, P_3 and P_4 have birthdays on consecutive dates is $\frac{6}{365^2}$, and so the probability that all four passengers have birthdays on three consecutive dates, with P_1 and P_2 having the same birthday, is $\frac{6}{365^3}$. Finally, the answer to the problem is obtained by multiplying this probability by 6, obtaining the result $\frac{36}{365^3}$.

Method 3

Another possible approach is along lines fairly similar to those followed in Method 1, but considers first the probability that the birthdays of the four passengers fall on three *different* dates, leaving consideration of the problem of *consecutive* dates until the end of the solution. As in the earlier solution, we consider the four passengers in order and, for $i = 2, 3, 4$, let D_i stand for the event that P_i's birthday is *different* from those considered earlier. The probability that the four birthdays fall on three different days is then

$$\Pr(D_2 \cap D_3 \cap \overline{D}_4) + \Pr(D_2 \cap \overline{D}_3 \cap D_4) + \Pr(\overline{D}_2 \cap D_3 \cap D_4)$$

$$= \left(\frac{364}{365} \times \frac{363}{365} \times \frac{3}{365}\right) + \left(\frac{364}{365} \times \frac{2}{365} \times \frac{363}{365}\right) + \left(\frac{1}{365} \times \frac{364}{365} \times \frac{363}{365}\right)$$

$$= \frac{6 \times 364 \times 363}{365^3}.$$

The probability we have just obtained is composed of equal components corresponding to each of the $\binom{365}{3}$ possible combinations of three days, but only 365 such combinations are of three *consecutive* days. The probability that the four have birthdays on three consecutive days is thus obtained by multiplying the probability obtained in the previous paragraph by

$$\frac{365}{\binom{365}{3}} = \frac{6}{364 \times 363};$$

we obtain the same probability as by the other two methods.

Method 4

We consider a sample space of outcomes of the form (b_1, b_2, b_3, b_4), where b_i denotes the birthday of passenger P_i: there are 365^4 such outcomes, all equally likely. To solve the problem we must count the number of outcomes having the required property. This can be done by considering a particular set of three days and then multiplying by 365.

For a particular set of three days, there are three possibilities to consider, depending on whether the duplicated birthday is on the first, second or third day. Consider one of these: the number of possible allocations of passengers to the three birthdays is $4!/2 = 12$. The total number of outcomes with the required property is therefore $365 \times 3 \times 12$ and, dividing by 365^4, we obtain the same result as before.

Notes

(1) In our solution we ignore leap years and do not take account of the fact that birthdays are not uniformly distributed throughout the year. To allow for these two factors would complicate matters immensely.

(2) The solution using Method 1 is, in a sense, the least elegant of those presented, but represents possibly the most reliable approach. The method depends on the systematic enumeration of all outcomes favourable to the required event, and we have chosen to do this by considering outcomes involving two 'adjacent' birthdays first, and then dealing with those involving a 'close' and a 'between' birthday.

There are other methods of enumerating all favourable outcomes. One is by means of a tree diagram, such as that sketched in Figure 1.1.

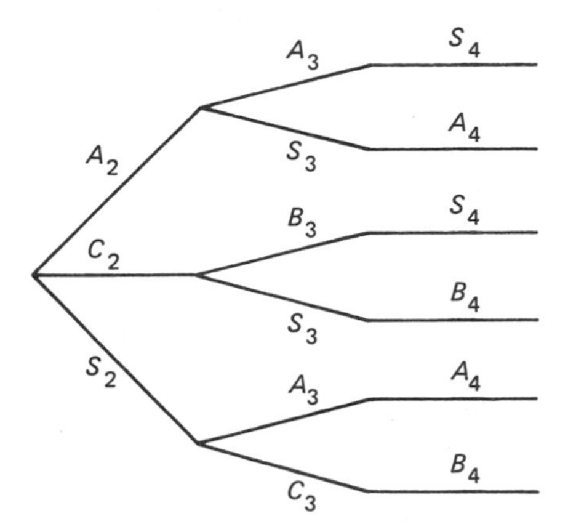

Figure 1.1 Tree diagram for events in Problem 1A.5

Construction of the diagram begins with the observation that, for the event of interest to occur, we must have one of A_2, C_2 and S_2 (B_2 being meaningless); if we have A_2, then we must have A_3 or S_3; and so on. (For further comments on the rôle of tree diagrams in solving probability problems, see Note 6 to Problem 1A.10.)

Another approach, closely related to that described above, is to consider all outcomes of the form $D_2 \cap E_3 \cap F_4$ (or, to use a suitable contracted notation, DEF), where each of D, E and F is one of A, B, C and S. There are 64 such outcomes, which we may imagine listed in some systematic way, for example in alphabetical order: AAA, AAB, AAC, . . . , SSC, SSD. Of the outcomes in this list some (such as the first) do not contribute to the event of interest, while others (such as the second) are meaningless or impossible. Eliminating these, we find that the outcomes remaining are AAS, ASA, CBS, CSB, SAA and SCB.

(3) Checking solutions to probability problems can be difficult – as we have remarked elsewhere, one is often uneasy until two different solutions to a problem yield the same answer. Since we have presented four solutions which agree with one another and with the odds quoted by the authors of the letter to *The Times*, we feel happy that our solution is correct. It is, however, only fair to confess that, while working on this problem, one of us also produced a number of incorrect solutions. It is also interesting to note that we were, at one stage, somewhat confused by the fact that a check that we applied to a correct solution seemed to fail.

It is sometimes useful to consider, for checking purposes, a small-scale version of a problem. In this case we considered the same problem for a 'year' of 3 days, since the entire sample space of 81 outcomes could then be easily enumerated. Unfortunately the methods we have used in our solution then lead to a probability of $\frac{36}{3^3}$, which is clearly ridiculous. On reflection, we realised that the arguments we have used all break down for such a short 'year': for example, in Method 1, a 'close' birthday is also an 'adjacent' birthday. They are, however, valid for a 'year' of 4 days, for which the probability is $\frac{36}{4^3}$.

(4) Letters of the the type quoted above appear, from time to time, in the press: generally the smallness of the probability seems to be taken as a 'surprise index', a measure of the remarkable

nature of the event.

It is instructive, however, to consider how this 'surprise index' ought to be assessed. If a letter were submitted to *The Times* to the effect that four people had birthdays on January 12, February 23, April 24 and September 6, one imagines that the editor might not find space for it – yet the probability of this event, $4!/365^4$, is far smaller than that referred to in this problem. To judge how surprised one ought to be by a claim, one must consider the probability, not only of the event itself, but also of any other event which would justify a similar claim. It is this fact which rules out a claim based on the four dates specified above. In the case of the four passengers, a letter might also have been sent in the event of

(i) all birthdays on at most two consecutive days,

(ii) three birthdays on one day,

(iii) all birthdays on the same day of the month (but in different months),

and so on. Even after allowing for this, however, it is likely that the observed event will still seem quite unusual.

To calculate a useful 'surprise index' one has not just to calculate the probability of the event observed, but to add the probabilities of all other events of the same type, but even more 'unusual'. This is, of course, the sort of calculation required in finding a significance level in hypothesis testing, and we shall return to this topic in Chapter 4.

1A.6 Winning at craps

In the game of craps, the player throws two dice. On his first throw he wins if the total is 7 or 11, and loses if it is 2, 3, or 12. If his total on the first throw is not one of these figures, he continues to throw the dice until he either repeats the total of his first throw (in which case he wins) or obtains a total of 7 (in which case he loses). Find the probability that he wins.

Solution

The probability distribution of the total score on two dice is obtained as the solution to Problem 1B.2. Dealing first with the possibility that the player wins on his first throw, we see that this has probability

$$\begin{aligned}\Pr(\text{score} = 7) + \Pr(\text{score} = 11) &= \frac{6}{36} + \frac{2}{36}\\ &= \frac{8}{36}.\end{aligned}$$

(Although it might seem natural to simplify the fractions involved here, it will be more convenient to retain the common denominator 36 in all probabilities until later in the solution.)

We now have to deal with the remaining possibilities. First, consider the case when the first throw results in a 4: this has probability $\frac{3}{36}$ which, for the moment, we shall denote by w. Similarly, the probability of a score of 7, which is $\frac{6}{36}$, will be denoted by l. The introduction of this notation will enable us to deal with other possibilities for the first throw merely by changing the values of w and l.

Given that the first throw results in a 4, the probability of a win at the ith subsequent throw ($i = 1, 2, \ldots$) is $(1 - w - l)^{i-1}w$ (since $i - 1$ throws which result in neither a 4 nor a 7 have to be followed by a final throw which results in a 4). Summing these probabilities, we find that the probability of a win, given that the first throw results in a score of 4, is

$$w\sum_{i=1}^{\infty}(1 - w - l)^{i-1} = \frac{w}{w + l}.$$

The contribution to the total probability of a win associated with the first score being a 4 is then

$$\Pr(\text{first score is 4}) \times \Pr(\text{win} \mid \text{first score is 4}) = \frac{w^2}{w+l} = \frac{\left(\frac{3}{36}\right)^2}{\frac{3}{36}+\frac{6}{36}} = \frac{1}{36} \times \frac{3^2}{3+6}.$$

Since the probability of a score of 10 is also $\frac{3}{36}$, the contribution to the total probability of winning associated with that initial score is the same as the above. We may deal similarly with initial scores of 5 and 9 (with $w = \frac{4}{36}$), and with initial scores of 6 and 8 (with $w = \frac{5}{36}$). Combining all these results, we obtain the probability of winning as

$$\frac{8}{36} + \frac{2}{36} \times \left(\frac{3^2}{3+6} + \frac{4^2}{4+6} + \frac{5^2}{5+6}\right) = \frac{8}{36} + \left(\frac{2}{36} \times \frac{4824}{990}\right) = \frac{244}{495}.$$

Notes

(1) It is interesting to see how close the probability of winning (0·493) is to one half. If the game is played in a casino, there is a small advantage to the house, so that it makes a steady profit; players, however, perceiving the odds to be close to even, are happy to play the game.

Instances of gambling odds very close to even have been the subject of interest from the earliest days of the mathematical study of probability theory. In the seventeenth century the Chevalier de Méré, a French gambler, approached the mathematician Pascal with a number of problems. One of these concerned his observation that, while he had a better than even chance of throwing at least one six in four throws of a die, his chance of throwing at least one double six in twenty-four throws of a pair of dice was rather less than even. The probabilities of success in these two ventures can be seen to be

$$1 - \left(\frac{5}{6}\right)^4 = 0{\cdot}518 \quad \text{and} \quad 1 - \left(\frac{35}{36}\right)^{24} = 0{\cdot}491$$

respectively.

* (2) The fact that the probability of obtaining a 4 before a 7 (or, more generally, of obtaining a winning score before a losing score) takes the simple form $\frac{w}{w+l}$, coupled with the observation that this is just the conditional probability of scoring 4, given that the score is either 4 or 7, suggests an alternative argument which does not involve the summation of series. The basis of this argument is that we condition on the number of throws until the game terminates, and use the law of total probability, forming a weighted average of the conditional probabilities obtained. Thus

$$\Pr(\text{player wins}) = \sum_{k=2}^{\infty} \Pr(\text{game terminates at throw } k) \times \Pr(\text{player wins} \mid \text{game terminates at throw } k).$$

(Although we have not indicated it explicitly, all probabilities in the above are conditional on the first throw resulting in a 4.)

At this stage it appears as if this method, like that used in the main solution, involves the summation of series, but we can avoid this when we note that the conditional probability of winning appearing on the right hand side does not depend on the value of k on which we are conditioning: it is

$$\Pr(\text{score is 4} \mid \text{score is 4 or 7}) = \frac{w}{w + l}.$$

The unconditional probability of winning must therefore have the same value. (Strictly speaking, the need to sum a series has not disappeared. But the right hand side of the above expression becomes

$$\frac{w}{w + l} \times \sum_{k=2}^{\infty} \Pr(\text{game terminates at throw } k) = \frac{w}{w + l} \times 1,$$

the sum being 1 since it is the sum of the probability function of the length of the game.)

1A.7 The fisherman

Each Sunday a fisherman visits one of three possible locations near his home: he goes to the sea with probability $\frac{1}{2}$, to a river with probability $\frac{1}{4}$, or to a lake with probability $\frac{1}{4}$. If he goes to the sea there is an 80% chance that he will catch fish; corresponding figures for the river and the lake are 40% and 60% respectively.

(a) Find the probability that, on a given Sunday, he catches fish.

(b) Find the probability that he catches fish on at least two of three consecutive Sundays.

(c) If, on a particular Sunday, he comes home without catching anything, where is it most likely that he has been?

(d) His friend, who also goes fishing every Sunday, chooses among the three locations with equal probabilities. Find the probability that the two fishermen will meet at least once in the next two weekends.

(Any assumptions you make in solving this problem should be clearly stated.)

Solution

We shall use the following notation:

S: he goes to the sea;
R: he goes to the river;
L: he goes to the lake;
F: he catches fish.

The given information may then be written as

$$\Pr(S) = \frac{1}{2}, \quad \Pr(F \mid S) = \frac{4}{5},$$

$$\Pr(R) = \frac{1}{4}, \quad \Pr(F \mid R) = \frac{2}{5},$$

$$\Pr(L) = \frac{1}{4}, \quad \Pr(F \mid L) = \frac{3}{5}.$$

(a) Using the law of total probability,

$$\begin{aligned}\Pr(F) &= \Pr(S)\Pr(F \mid S) + \Pr(R)\Pr(F \mid R) + \Pr(L)\Pr(F \mid L) \\ &= \left(\frac{1}{2} \times \frac{4}{5}\right) + \left(\frac{1}{4} \times \frac{2}{5}\right) + \left(\frac{1}{4} \times \frac{3}{5}\right) \\ &= \frac{13}{20}.\end{aligned}$$

(b) It follows from the solution to part (a) that the number of Sundays on which he catches fish is a random variable X having the binomial distribution $B(3, \frac{13}{20})$. The required probability is thus

$$\Pr(X = 2) + \Pr(X = 3) = 3\left(\frac{13}{20}\right)^2\left(1 - \frac{13}{20}\right) + \left(\frac{13}{20}\right)^3$$

$$= \frac{2873}{4000}.$$

(c) From (a), $\Pr(\bar{F}) = 1 - \Pr(F) = \frac{7}{20}$. Hence

$$\Pr(S \mid \bar{F}) = \frac{\Pr(S \cap \bar{F})}{\Pr(\bar{F})}$$

$$= \frac{\Pr(S)\Pr(\bar{F} \mid S)}{\Pr(\bar{F})}$$

$$= \frac{\frac{1}{2}\left(1 - \frac{4}{5}\right)}{\frac{7}{20}} = \frac{2}{7}.$$

Similarly

$$\Pr(R \mid \bar{F}) = \frac{\frac{1}{4}\left(1 - \frac{2}{5}\right)}{\frac{7}{20}} = \frac{3}{7};$$

$$\Pr(L \mid \bar{F}) = \frac{\frac{1}{4}\left(1 - \frac{3}{5}\right)}{\frac{7}{20}} = \frac{2}{7}.$$

So it is most likely that he has been to the river.

(d) If, now, we let S_1 denote the event that the first fisherman goes to the sea and S_2 the event that the second goes to the sea, and define R_1, R_2, L_1, L_2 similarly, the probability that they meet on a given Sunday is

$$\Pr(S_1 \cap S_2) + \Pr(R_1 \cap R_2) + \Pr(L_1 \cap L_2)$$

$$= \Pr(S_1)\Pr(S_2) + \Pr(R_1)\Pr(R_2) + \Pr(L_1)\Pr(L_2)$$

$$= \left(\frac{1}{2} \times \frac{1}{3}\right) + \left(\frac{1}{4} \times \frac{1}{3}\right) + \left(\frac{1}{4} \times \frac{1}{3}\right)$$

$$= \frac{1}{3}.$$

The probability that they fail to meet over two weekends is then $(1 - \frac{1}{3})^2 = \frac{4}{9}$, so that the probability that they meet at least once is $1 - \frac{4}{9} = \frac{5}{9}$.

The assumptions made in solving this problem are, in parts (b) and (d), that the fishermen choose independently in different weeks and, in part (d), that they choose independently of one another. (There is a further assumption in part (d), namely that they meet if they both go to the same location.)

Notes

(1) In (c) we are using Bayes' Theorem, although this is, to some extent, disguised by the fact that the denominator, $\Pr(\overline{F})$, is obtained from the answer to part (a).

(2) We can simplify the answer to (d) by observing that the probability that they meet on a given Sunday is just the probability that the second chooses the same location as the first, namely $\frac{1}{3}$. This argument applies only since the second chooses each of the three locations with the same probability.

1A.8 Testing lightbulbs

The stock of a warehouse consists of boxes of high, medium and low quality lightbulbs in respective proportions 1:2:2. The probabilities of bulbs of the three types being unsatisfactory are 0·0, 0·1 and 0·2 respectively. If a box is chosen at random and two bulbs in it are tested and found to be satisfactory, what is the probability that it contains bulbs

(i) of high quality;

(ii) of medium quality;

(iii) of low quality?

Solution

We shall adopt the following notation for events of interest:

H: the box chosen contains high quality bulbs;
M: the box chosen contains medium quality bulbs;
L: the box chosen contains low quality bulbs;
S: the two bulbs tested are found to be satisfactory.

The given information concerning the proportion of boxes of the three types may be written in the following form: $\Pr(H) = 0{\cdot}2$, $\Pr(M) = 0{\cdot}4$, $\Pr(L) = 0{\cdot}4$. From the information on quality, we deduce that

$$\Pr(S \mid H) = 1{\cdot}0,$$
$$\Pr(S \mid M) = (1{\cdot}0 - 0{\cdot}1)^2 = 0{\cdot}81,$$
$$\Pr(S \mid L) = (1{\cdot}0 - 0{\cdot}2)^2 = 0{\cdot}64.$$

We may now use Bayes' Theorem. Probability (i) is

$$\Pr(H \mid S) = \frac{\Pr(H)\Pr(S \mid H)}{\Pr(H)\Pr(S \mid H) + \Pr(M)\Pr(S \mid M) + \Pr(L)\Pr(S \mid L)}$$
$$= \frac{0{\cdot}2 \times 1{\cdot}0}{(0{\cdot}2 \times 1{\cdot}0) + (0{\cdot}4 \times 0{\cdot}81) + (0{\cdot}4 \times 0{\cdot}64)}$$
$$= \frac{0{\cdot}2}{0{\cdot}2 + 0{\cdot}324 + 0{\cdot}256}$$
$$= \frac{0{\cdot}2}{0{\cdot}78} = 0{\cdot}256.$$

Probabilities (ii) and (iii) are obtained similarly. For (ii), we find

$$\Pr(M \mid S) = \frac{0{\cdot}324}{0{\cdot}78} = 0{\cdot}415$$

and, for (iii),

$$\Pr(L \mid S) = \frac{0{\cdot}256}{0{\cdot}78} = 0{\cdot}328.$$

Notes

(1) This is a fairly standard type of problem involving the use of Bayes' Theorem, made slightly more complicated by the fact that *two* bulbs are being tested. Note the computational simplification which results from writing down the values (0·2, 0·324 and 0·256) of the probabilities $\Pr(H \mid S)$, $\Pr(M \mid S)$ and $\Pr(L \mid S)$ in the course of finding probability (i): probabilities (ii) and (iii) then follow almost immediately, without the need to 'start afresh'.

(2) The three probabilities obtained should, of course, add to 1·0. The fact that their total appears to be 0·999 is due to the rounding errors created by recording the calculated values to three decimal places.

1A.9 Designing a navigation system

A spacecraft navigation system consists of components of two types – A and B – connected in a series-parallel arrangement, as in Figure 1.2.

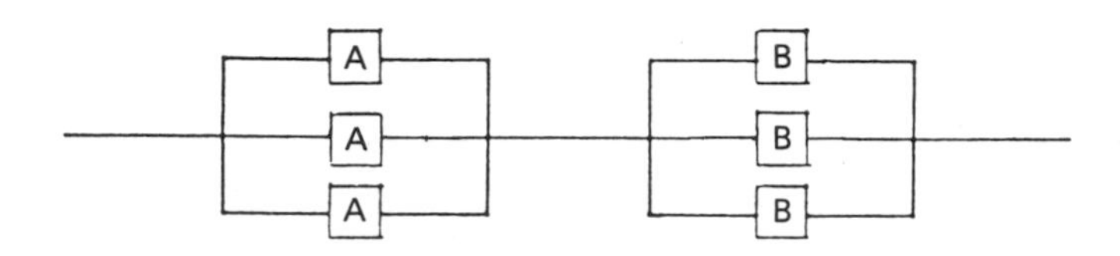

Figure 1.2 Illustration of series-parallel system

The system will function as long as at least one component of each type functions. The probability that a type A component will function correctly throughout the planned lifetime of the spacecraft is 0·9; the corresponding probability for a component of type B is 0·8. Components may be assumed to fail independently of one another.

The numbers of components of the two types may be varied in order to adjust the probability of failure of the navigation system.

(a) Show that a requirement that the probability of failure of the entire system should be less than 1% may be met by using three components of each type (as in the diagram above).

(b) If components of type A weigh 1 kg, while those of type B weigh 2 kg, what is the composition of the most reliable navigation system weighing 11 kg or less?

Solution

(a) Suppose the system consists of a components of type A and b components of type B. Then

$$\begin{aligned}
\Pr(\text{system functions}) &= \Pr(\text{at least one type A component functions}) \\
&\quad \times \Pr(\text{at least one type B component functions}) \\
&= \{1{\cdot}0 - \Pr(\text{all type A components fail})\} \\
&\quad \times \{1{\cdot}0 - \Pr(\text{all type B components fail})\} \\
&= (1{\cdot}0 - 0{\cdot}1^a) \times (1{\cdot}0 - 0{\cdot}2^b).
\end{aligned}$$

If $a = b = 3$, this equals $0{\cdot}999 \times 0{\cdot}992 = 0{\cdot}991$ and, since this exceeds 0·99, the requirement of part (a) is met.

(b) In order to solve this part, we could examine all systems weighing 11 kg (there is clearly no point in considering systems weighing less, since any such system can be improved upon by adding components of type A). There are five systems in all; we can, however, reduce the number of possibilities that we have to investigate if we observe that the system considered in part (a) weighs less than 11 kg and, therefore, that the system we are seeking must exceed this in reliability, i.e. must have reliability exceeding 0·991. It follows that a and b must satisfy

$$1{\cdot}0 - 0{\cdot}1^a \geq 0{\cdot}991, \text{ i.e. } a \geq 3,$$

and

$$1{\cdot}0 - 0{\cdot}2^b \geq 0{\cdot}991, \text{ i.e. } b \geq 3.$$

This leaves just two possibilities to be considered: $a = 5$, $b = 3$, with reliability

$$(1{\cdot}0 - 0{\cdot}1^5)\times(1{\cdot}0 - 0{\cdot}2^3) = 0{\cdot}992,$$

and $a = 3$, $b = 4$, with reliability

$$(1{\cdot}0 - 0{\cdot}1^3)\times(1{\cdot}0 - 0{\cdot}2^4) = 0{\cdot}997,$$

so that the best configuration weighing 11 kg or less consists of three components of type A and four of type B.

Note

Part (b) is an example of a *knapsack* problem, in which integer variables (here a and b) must be chosen to maximise some quantity (here the system reliability) while satisfying constraints imposed by available resources (weight in this case). Although small-scale problems of this type can be solved relatively easily, since there are relatively few possibilities to consider, the solution of larger problems requires a more sophisticated approach.

1A.10 The sports club

Three quarters of the members of a sports club are adults, and one quarter are children. Three quarters of the adults, and three fifths of the children, are male. Half the adult males, and a third of the adult females, use the swimming pool at the club; the corresponding proportion for children of either sex is four fifths.

(a) Find the probability that a member of the club uses the swimming pool.

(b) Find the probability that a member of the club who uses the swimming pool is male.

(c) Find the probability that a member of the club is female.

(d) Find the probability that a member of the club who uses the swimming pool is female.

(e) Find the probability that a male user of the swimming pool is a child.

(f) Find the probability that a member of the club who does not use the swimming pool is either female or an adult.

Solution

In our solution we shall let A denote the event that a member of the club is an adult, and C the event that he or she is a child. We shall let M and F stand, respectively, for the events that a member is male or female, and S for the event that he or she uses the swimming pool. The given information may then be summarised as follows:

$$\Pr(A) = \frac{3}{4}, \quad \Pr(C) = \frac{1}{4};$$

$$\Pr(M \mid A) = \frac{3}{4}, \quad \Pr(F \mid A) = \frac{1}{4};$$

$$\Pr(M \mid C) = \frac{3}{5},\quad \Pr(F \mid C) = \frac{2}{5};$$

$$\Pr(S \mid A \cap M) = \frac{1}{2},\quad \Pr(S \mid A \cap F) = \frac{1}{3};$$

$$\Pr(S \mid C \cap M) = \Pr(S \mid C \cap F) = \Pr(S \mid C) = \frac{4}{5}.$$

(a)
$$\begin{aligned}\Pr(S) &= \Pr(A \cap M \cap S) + \Pr(A \cap F \cap S) + \Pr(C \cap S)\\ &= \Pr(A)\Pr(M \mid A)\Pr(S \mid A \cap M) + \Pr(A)\Pr(F \mid A)\Pr(S \mid A \cap F)\\ &\quad + \Pr(C)\Pr(S \mid C)\\ &= \left(\frac{3}{4}\times\frac{3}{4}\times\frac{1}{2}\right) + \left(\frac{3}{4}\times\frac{1}{4}\times\frac{1}{3}\right) + \left(\frac{1}{4}\times\frac{4}{5}\right)\\ &= \frac{9}{32} + \frac{1}{16} + \frac{1}{5}\\ &= \frac{87}{160} = 0{\cdot}544.\end{aligned}$$

(b)
$$\Pr(M \mid S) = \frac{\Pr(M \cap S)}{\Pr(S)}.$$

Now $\Pr(S)$ has been obtained as the answer to part (a), and

$$\begin{aligned}\Pr(M \cap S) &= \Pr(A \cap M \cap S) + \Pr(C \cap M \cap S)\\ &= \frac{9}{32} + \Pr(C)\Pr(M \mid C)\Pr(S \mid M \cap C)\\ &= \frac{9}{32} + \left(\frac{1}{4}\times\frac{3}{5}\times\frac{4}{5}\right)\\ &= \frac{9}{32} + \frac{3}{25} = \frac{321}{800}.\end{aligned}$$

So

$$\Pr(M \mid S) = \frac{\frac{321}{800}}{\frac{87}{160}} = 0{\cdot}738.$$

(c)
$$\begin{aligned}\Pr(F) &= \Pr(A)\Pr(F \mid A) + \Pr(C)\Pr(F \mid C)\\ &= \left(\frac{3}{4}\times\frac{1}{4}\right) + \left(\frac{1}{4}\times\frac{2}{5}\right)\\ &= \frac{3}{16} + \frac{1}{10} = 0{\cdot}288.\end{aligned}$$

(d) We could proceed here as in the solution to part (b), but can save work by making use of the solution to that part of the question, since the required probability is simply

$$\Pr(F \mid S) = 1 - \Pr(M \mid S).$$

Since $\Pr(M \mid S) = 0{\cdot}738$ the solution is $0{\cdot}262$.

(e) $$\Pr(C \mid M \cap S) = \frac{\Pr(C \cap M \cap S)}{\Pr(M \cap S)}.$$

The numerator and denominator in the above expression have already appeared in the solution to part (b). Substituting their values, we obtain

$$\Pr(C \mid M \cap S) = \frac{\frac{3}{25}}{\frac{321}{800}} = 0{\cdot}299.$$

(f) The required probability is

$$\Pr(A \cup F \mid \bar{S}) = \frac{\Pr\{(A \cup F) \cap \bar{S}\}}{\Pr(\bar{S})}.$$

The denominator in this expression is easily obtained from the solution to part (a):

$$\Pr(\bar{S}) = 1 - \Pr(S) = \frac{73}{160}.$$

To obtain the value of the numerator, it is simplest to split the event of interest into three mutually exclusive components, and to sum their probabilities. We obtain

$$\begin{aligned}\Pr\{(A \cup F) \cap \bar{S}\} &= \Pr(A \cap M \cap \bar{S}) + \Pr(A \cap F \cap \bar{S}) + \Pr(C \cap F \cap \bar{S}) \\ &= \Pr(A)\Pr(M \mid A)\Pr(\bar{S} \mid A \cap M) + \Pr(A)\Pr(F \mid A)\Pr(\bar{S} \mid A \cap F) \\ &\quad + \Pr(C)\Pr(F \mid C)\Pr(\bar{S} \mid C \cap F) \\ &= \left(\frac{3}{4} \times \frac{3}{4} \times \frac{1}{2}\right) + \left(\frac{3}{4} \times \frac{1}{4} \times \frac{2}{3}\right) + \left(\frac{1}{4} \times \frac{2}{5} \times \frac{1}{5}\right) \\ &= \frac{9}{32} + \frac{1}{8} + \frac{1}{50} = \frac{341}{800}.\end{aligned}$$

We now obtain the required probability:

$$\Pr(A \cup F \mid \bar{S}) = \frac{\frac{341}{800}}{\frac{73}{160}} = 0{\cdot}934.$$

Notes

(1) For the sake of accuracy, we have obtained the answer to each part in fractional form, converting to a decimal at the end. The disadvantages of earlier conversion may be illustrated by considering part (b): using $\Pr(M \cap S) = 0{\cdot}401$ and $\Pr(S) = 0{\cdot}544$, we obtain the inaccurate result $\Pr(M \mid S) = 0{\cdot}737$.

(2) Although the statement of this problem does not include the words 'randomly selected', our answer assumes, in effect, that these words have been inserted after the words 'Find the probability that a' in each part. It is not uncommon to encounter examination questions stated in a somewhat loose manner, and to have to make suitable assumptions in order to reach a solution. In this problem the assumption of random selection is the only reasonable one to make. (As discussed in Note 1 to Problem 1A.1, the word 'random' has a rather specific meaning here.) The assumption might not be valid in real life: consider, for example, answering part (a) when the answer is to relate to a member visiting the club on a morning when the swimming pool is closed.

(3) In our solutions to later parts we have, where it helps to shorten the solution, made use of results appearing in the solutions to earlier parts. There is, however, no need to do this and, in an earlier draft of the solution, we did so to a lesser extent than we do now. In solving examination questions, it is important to bear in mind that work done in solving earlier parts of a question may pave the way for solving later parts. It is important, however, to guard against spending too much time looking for such 'short-cuts' or attempting to use them at all costs.

Sometimes, even if earlier results *can* be used, it is easier to start afresh. This is the case in part (f) of the present problem.

(4) There is a simpler solution to part (f). Rather than find the required conditional probability directly, we seek the conditional probability of the complementary event (that the member is a male child). This is

$$\Pr(C \cap M \mid \overline{S}) = \frac{\Pr(C \cap M \cap \overline{S})}{\Pr(\overline{S})}.$$

Now the denominator in this expression is $\frac{73}{160}$, as before; and the numerator is

$$\Pr(C)\Pr(M \mid C)\Pr(\overline{S} \mid C \cap M) = \frac{1}{4} \times \frac{3}{5} \times \frac{1}{5} = \frac{3}{100}.$$

Substituting these values we obtain

$$\Pr(C \cap M \mid \overline{S}) = 0{\cdot}066,$$

and hence

$$\Pr(A \cup F \mid \overline{S}) = 1 - 0{\cdot}066 = 0{\cdot}934,$$

as obtained before.

(5) This is a rather unappealing problem, not because the concepts involved are particularly advanced, but because of the complicated nature of the manipulations required and, possibly, because it is difficult to imagine why anyone would be interested in the more intricate questions asked. It is, however, not unlike some examination questions that we have seen – hence its inclusion.

The solution we have presented is fairly concise, due to the use of set-theoretic notation; but, like many arguments which are, apparently, of a fairly abstract mathematical nature, it can be understood more easily by bearing in mind, while studying it, the meaning of the various events involved. The arguments could be presented verbally, but the result would become very lengthy. Thus, for example, we could begin the solution to part (a) in the following manner.

> 'A swimmer can be either an adult or a child, so we have to add the probabilities of an adult swimmer and a child swimmer. We need not take the sex of the child into account in finding the latter probability, since the probability that a child is a swimmer does not depend on sex, but we do have to take account of sex in finding the probability of an adult swimmer: this probability is then expressed as the sum of the probabilities of an adult male swimmer and an adult female swimmer . . . '

So far, we've dealt with the first line of our solution, and there's a lot more to go! The advantages of our more concise approach in *presenting* a solution are evident, but it is important to bear in mind in *reading* the solution (particularly if it seems rather mystifying) that it represents an argument along the lines we have expanded in words above for the initial section, and mentally to perform similar expansions on other sections. (See Note 1 to Problem 1A.2 and Note 1 to Problem 1B.3 for further discussion of the rôle of set theoretic notation in the solution of problems in probability.)

(6) One way of getting to grips with a problem such as this is by means of a rough tree diagram. Readers will notice that we do not make much use of tree diagrams in this chapter. This reflects our view that, while tree diagrams have undoubted advantages in the teaching of probability and in the early stages of solving a complicated problem, they do not always constitute the best method of presenting the final solution. A tree diagram on its own is not enough. It has to be supplemented by sufficient verbal explanation to establish conventions, define notation, and to relate the diagram to the problem and its solution. With many problems it turns out, after a tree diagram has been produced, that only a small section of the diagram is needed for a solution. For these reasons we have, in solving a number of problems for which a tree diagram might have been considered, preferred to present a verbal solution (we include in this category solutions in which many of the 'words' are mathematical symbols).

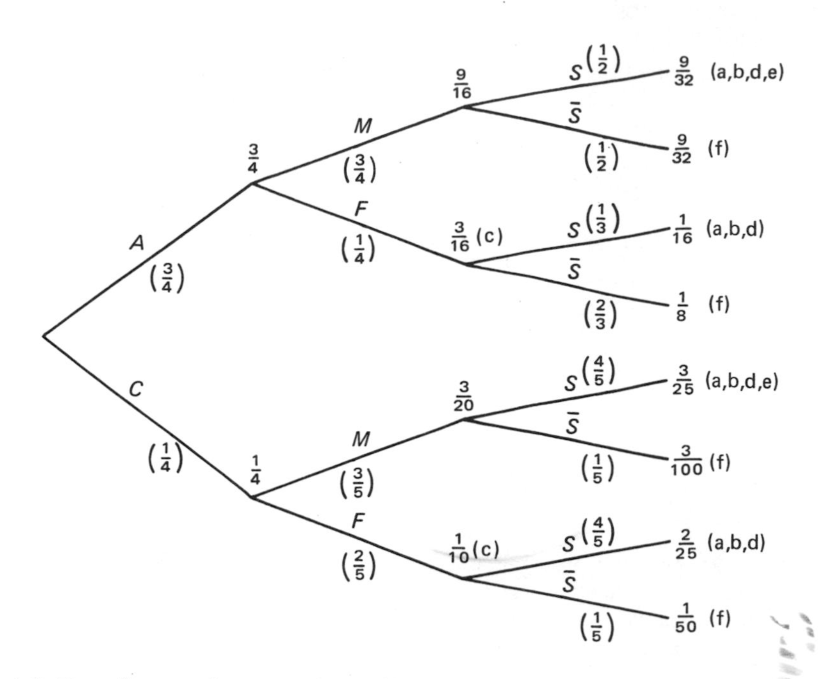

Figure 1.3 Tree diagram for events in Problem 1A.10

We present, in Figure 1.3, a tree diagram relating to this problem. The notation adopted is that already described, and the conventions followed are that the probabilities in brackets alongside the branches of the tree are appropriate conditional probabilities, whereas those appearing, unbracketed, at the nodes are the unconditional probabilities of the corresponding combinations of events. A solution of the problem based on this diagram would be completed by identifying, for each part, the probabilities to be extracted from the diagram, and performing, with appropriate explanation, the necessary operations on these probabilities. Since this would repeat much of what we have already done, we shall not do this. We have, however, indicated on the diagram where the probabilities required for each part can be found, and the reader can thus relate the diagram to our earlier solution. It will be noted that most of the probabilities on the diagram are used in solving at least one part of the problem, so that not much of the work involved in the construction of the diagram is 'wasted'. The situation would have been somewhat different for a problem consisting of only a few of the six parts of the present problem, and it is interesting to note that, while our original solution to part (a) involves the addition of three probabilities, a solution based on a tree diagram would involve the addition of four and, quite possibly, the calculation of several others.

1A.11 Playing the fruit machine

A fruit machine takes 10p coins. At each play it pays out 30p with probability $\frac{1}{4}$; no other payouts are possible.

(a) Find the probability that a player who starts with 20p and plays until her funds are exhausted plays at least six times.

(b) Find the probability that a player who still has some money left after playing five times has at least three more plays.

Solution

(a) We shall represent possible sequences of results by sequences of letters, letting W stand for a win and L for a loss. To find the probability of at least six plays, we shall find the probability of five or fewer and subtract from 1. There are three possibilities to be considered: LL, LWLLL, and WLLLL. Assuming independence of successive plays, the corresponding probabilities are

$$\left(\frac{3}{4}\right)^2, \ \frac{1}{4}\times\left(\frac{3}{4}\right)^4, \text{ and } \ \frac{1}{4}\times\left(\frac{3}{4}\right)^4.$$

Thus the probability that she plays fewer than six times is

$$\left(\frac{3}{4}\right)^2 + 2\times\frac{1}{4}\times\left(\frac{3}{4}\right)^4 = 0{\cdot}721.$$

It follows that the probability that she plays at least six times is 0·279.

(b) What we require is

$$\begin{aligned}&\Pr(\text{she plays at least nine times} \mid \text{she plays at least six times})\\ &= \frac{\Pr(\text{she plays at least nine times})}{\Pr(\text{she plays at least six times})}.\end{aligned}$$

The denominator has been obtained already. To obtain the numerator, we proceed as before, finding the probability that she plays at most eight times and subtracting from 1. In addition to the three possibilities considered above, there are seven other possibilities which result in eight or fewer plays in all. These are LWWLLLLL, LWLWLLLL, LWLLWLLL, WWLLLLLL, WLWLLLLL, WLLWLLLL, and WLLLWLLL. The probability of eight or fewer plays in all is therefore

$$\left(\frac{3}{4}\right)^2 + 2\times\frac{1}{4}\times\left(\frac{3}{4}\right)^4 + 7\times\left(\frac{1}{4}\right)^2\times\left(\frac{3}{4}\right)^6 = 0{\cdot}799.$$

The required conditional probability is thus

$$\frac{1{\cdot}0 - 0{\cdot}799}{0{\cdot}279} = 0{\cdot}721.$$

Note

The numerator in the conditional probability obtained in the solution to part (b) should be, strictly speaking, the probability that the player plays at least six times *and* plays at least nine times. Since the latter implies the former, the probability reduces to that of the latter.

1B Random Variables

In this section we move on to problems in which the sample space consists of the integers, or the real line, or a subset of either. In such cases the experiment will have an outcome which is a number, which varies from one repetition of the experiment to another, and is called, naturally enough, a *random variable*.

The problems in this section concentrate on the distributions of random variables and their properties. Methods of investigating these differ according to the type of random variable; the most common types are *discrete* and *continuous* random variables. Both can have their distribution described by a *cumulative distribution function* (sometimes abbreviated to *distribution function* or *c.d.f.*); other functions which will be examined are the *probability function* (or *probability mass function*, or *p.f.*) for discrete random variables, and the *probability density function* (or *density function*, or *p.d.f.*) for continuous random variables.

The problems here also deal with the most important summarising measures for distributions based on the concept of *expectation*. In particular one is often required to calculate the *mean* and *variance* of a distribution.

The notation used will extend that used in Section 1A. Random variables are commonly denoted by capital letters, typically those near the end of the alphabet. So we might wish to calculate $\Pr(X = 3)$ or $\Pr(X \geq -2)$, or, more generally, $\Pr(X = x)$, where X denotes a random variable and x denotes a value which X might take. The cumulative distribution function of a random variable X will be denoted by $F_X(x)$, i.e. we define $F_X(x) = \Pr(X \leq x)$, as a function of x. If X is discrete, its probability function is $\Pr(X = x)$, and it is sometimes convenient to denote this by $p_X(x)$. If X is a continuous random variable, its probability density function will be denoted by $f_X(x)$. Note that it is the subscript which indicates the identity of the random variable; occasionally the subscript will be omitted when the context makes it unnecessary.

Our notation for expectation and variance is also natural and conventional. We denote the expectation of a random variable X simply by $\mathrm{E}(X)$, whether X be discrete or continuous. The concept of expectation extends to functions of random variables; the expectation of $g(X)$ will be denoted by $\mathrm{E}[g(X)]$. (The shape of any brackets used carries no significance.) In particular, for the special case $g(X) = \{X - \mathrm{E}(X)\}^2$, we obtain the variance of X, denoted here by $\mathrm{Var}(X)$. Thus

$$\mathrm{Var}(X) = \mathrm{E}[\{X - \mathrm{E}(X)\}^2].$$

Finally, we note that the reader may find problems in later chapters which are closely related to those in this section. Besides those in the following chapter, Problems 4D.1, 4D.2 and 5D.2 are worth mentioning in this connection.

1B.1 The highest score in a game of dice

(a) A random variable X is defined as the larger of the scores obtained in two throws of an unbiased, six-sided, die. Show that

$$\Pr(X = x) = \frac{(2x - 1)}{36}, \qquad x = 1, 2, \ldots, 6.$$

(b) A random variable Y is defined as the highest score obtained in k independent throws of an unbiased, six-sided, die. Find an expression for the probability function of Y.

Solution

(a) Let S_1 and S_2 denote the two scores, so that $X = \max(S_1, S_2)$. The two events $\{X \leq x\}$ and $\{S_1 \leq x\} \cap \{S_2 \leq x\}$ are clearly equivalent. Therefore

$$\begin{aligned} \Pr(X \leq x) &= \Pr(S_1 \leq x \cap S_2 \leq x) \\ &= \Pr(S_1 \leq x)\Pr(S_2 \leq x), \quad \text{by independence.} \end{aligned}$$

Now $\Pr(S_i = j) = \dfrac{1}{6}$, $i = 1, 2$; $j = 1, 2, \ldots, 6$, so that $\Pr(S_i \leq x) = \dfrac{x}{6}$, and therefore

$$\Pr(X \leq x) = \frac{x^2}{36}, \qquad x = 0, 1, 2, \ldots, 6.$$

But

$$\begin{aligned} \Pr(X = x) &= \Pr(X \leq x) - \Pr(X \leq x - 1), \\ &= \frac{1}{36}\{x^2 - (x-1)^2\} = \frac{(2x - 1)}{36}, \qquad x = 1, 2, \ldots, 6. \end{aligned}$$

(b) Let $S_1, \ldots, S_k$ denote the k scores, so that $Y = \max(S_1, \ldots, S_k)$. By an extension of the result above for the case $k = 2$, we obtain

$$\Pr(Y \le y) = \prod_{i=1}^{k} \Pr(S_i \le y) = \left(\frac{y}{6}\right)^k,$$

and so

$$\Pr(Y = y) = \Pr(Y \le y) - \Pr(Y \le y-1) = \frac{1}{6^k}\{y^k - (y-1)^k\}, \qquad y = 1, 2, \ldots, 6.$$

Notes

(1) An alternative solution to part (a) is as follows. The event $\{X = x\}$, that the *larger* of S_1 and S_2 equals x, occurs if $S_1 = x$ and $S_2 \le x$ or if $S_1 \le x$ and $S_2 = x$. Denoting these two events by A and B, we see that the event $\{X = x\}$ is just $A \cup B$, so we obtain

$$\Pr(X = x) = \Pr(A \cup B) = \Pr(A) + \Pr(B) - \Pr(A \cap B).$$

But $\Pr(A) = \frac{1}{6} \times \frac{x}{6}$, and by symmetry this is also $\Pr(B)$. Further, the event $A \cap B$ is just the event that $S_1 = x$ and $S_2 = x$, and so has probability 1/36. We thus find

$$\Pr(X = x) = 2 \times \frac{1}{6} \times \frac{x}{6} - \frac{1}{36} = \frac{2x-1}{36},$$

as required. While this method is more direct, and therefore seems more natural, it does not extend easily from the case $k = 2$ to the general case of part (b).

(2) In problems dealing with the largest of a set of random variables, it is often much simpler to find the cumulative distribution function than to find the probability function. (The latter can, of course, be calculated once the cumulative distribution function is known.) A similar approach may be adopted for the smallest of a set of random variables.

(3) The cumulative distribution function is also particularly useful when one needs to find the distribution of a random variable defined as a function of another. This technique is discussed in Problems 1B.10 and 1B.11.

(4) In part (b) we would expect that, as $k \to \infty$, $\Pr(Y = 6) \to 1$. We see that

$$\Pr(Y = y) = \frac{y^k}{6^k}\left\{1 - \left(\frac{y-1}{y}\right)^k\right\}.$$

So, as $k \to \infty$, $\Pr(Y = y) \to 0$ for $y < 6$, while $\Pr(Y = 6) \to 1$.

1B.2 The total score from two dice

Two unbiased dice are thrown simultaneously, and the sum of the scores on their uppermost faces is recorded. What is the distribution of this quantity, and what are its mean and variance?

Solution

The simplest approach is to denote the scores on the faces shown by the two dice by X and Y, and to write $Z = X + Y$ as the total, whose distribution is required. Clearly, the range of the random variable Z is the set of integers from 2 to 12 inclusive, and the probability distribution of Z is thus found by obtaining the set of probabilities $\Pr(Z = z)$, $z = 2, 3, \ldots, 12$. Starting with the event $Z = 2$, we see by independence that

$$\Pr(Z = 2) = \Pr(X = 1 \cap Y = 1)$$

$$= \Pr(X = 1) \times \Pr(Y = 1) = \frac{1}{6} \times \frac{1}{6} = \frac{1}{36},$$

and by essentially the same reasoning

$$\Pr(Z = 12) = \frac{1}{36}.$$

Now the event $Z = 3$ will occur if $X = 1$ and $Y = 2$ or if $X = 2$ and $Y = 1$, two mutually exclusive events. So, by the addition law of probability,

$$\Pr(Z = 3) = \{\Pr(X = 1) \times \Pr(Y = 2)\} + \{\Pr(X = 2) \times \Pr(Y = 1)\},$$

with the products appearing because events relating to X are independent of those relating to Y. Thus

$$\Pr(Z = 3) = \frac{1}{36} + \frac{1}{36} = \frac{2}{36},$$

and, again, similar reasoning shows that

$$\Pr(Z = 11) = \frac{2}{36}.$$

Extending the argument further, we find three possible patterns of values – in an obvious notation, (1, 3), (2, 2) and (3, 1) – contributing to the event $Z = 4$, and three patterns contributing to $Z = 10$, so we obtain

$$\Pr(Z = 4) = \Pr(Z = 10) = \frac{3}{36};$$

similarly

$$\Pr(Z = 5) = \Pr(Z = 9) = \frac{4}{36}.$$

Also

$$\Pr(Z = 6) = \Pr(Z = 8) = \frac{5}{36},$$

since for each of these there are five patterns of values of X and Y contributing to the event concerned.

Finally we show in detail the calculation for $\Pr(Z = 7)$, and note in passing that the other cases could also have been dealt with in this more rigorous way. We see that the event $Z = 7$ occurs when, whatever value x is taken by X, Y takes the value $7 - x$. So we obtain

$$\Pr(Z = 7) = \sum_{x=1}^{6} \Pr(X = x \cap Y = 7 - x)$$

$$= \sum_{x=1}^{6} \Pr(X = x) \Pr(Y = 7 - x), \text{ by independence,}$$

$$= \sum_{x=1}^{6} \frac{1}{6} \times \frac{1}{6} = \frac{6}{36}.$$

We thus obtain the first result we require, the probability distribution of the random variable Z, which can most conveniently be expressed as a table.

z	2	3	4	5	6	7	8	9	10	11	12
$\Pr(Z=z)$	$\frac{1}{36}$	$\frac{2}{36}$	$\frac{3}{36}$	$\frac{4}{36}$	$\frac{5}{36}$	$\frac{6}{36}$	$\frac{5}{36}$	$\frac{4}{36}$	$\frac{3}{36}$	$\frac{2}{36}$	$\frac{1}{36}$

We now require the mean and variance of the distribution of Z. The mean can in fact be found easily, by symmetry, but in principle the calculation is as follows.

$$E(Z) = \sum_{z=2}^{12} z \Pr(Z=z)$$

$$= \frac{1}{36}\{(2\times1) + (3\times2) + (4\times3) + (5\times4) + (6\times5) +$$

$$+ (7\times6) + (8\times5) + (9\times4) + (10\times3) + (11\times2) + (12\times1)\}$$

$$= \frac{252}{36} = 7.$$

To obtain the variance of Z we first obtain $E(Z^2)$, viz.

$$E(Z^2) = \frac{1}{36}\{(2^2\times1) + (3^2\times2) + (4^2\times3) + (5^2\times4) + (6^2\times5) +$$

$$+ (7^2\times6) + (8^2\times5) + (9^2\times4) + (10^2\times3) + (11^2\times2) + (12^2\times1)\}$$

$$= \frac{1974}{36}.$$

We now obtain the variance of Z simply, as

$$\mathrm{Var}(Z) = \frac{1974}{36} - 7^2 = \frac{35}{6}.$$

Hence $E(Z) = 7$ and $\mathrm{Var}(Z) = \frac{35}{6} = 5{\cdot}833$.

Notes

(1) The method used in the solution to obtain the variance of Z is, generally, the best one to use, but occasionally some feature of a problem makes another approach slightly easier, and in fact this is the case here. There are two alternative methods, both of which are slightly quicker than that used in the solution. The first of these uses the fact that the distribution of Z is symmetrical about $E(Z)$, which is itself a convenient integer, so that $\mathrm{Var}(Z)$ can be calculated easily from the formula usually used only as a definition, viz.

$$\mathrm{Var}(Z) = E[\{Z - E(Z)\}^2].$$

Noting that $\{Z-7\}^2$ takes only 6 values, 0, 1, 4, 9, 16 and 25, we obtain

$$\mathrm{Var}(Z) = \frac{1}{36}\{(6\times0) + (10\times1) + (8\times4) + (6\times9) + (4\times16) + (2\times25)\} = \frac{35}{6},$$

as before.

The other method is more sophisticated, but can be extended readily. We recall that Z is the sum of two random variables X and Y, so that

$$E(Z) = E(X) + E(Y),$$

and since X and Y are independent we also have the result

$$\mathrm{Var}(Z) = \mathrm{Var}(X) + \mathrm{Var}(Y) = 2\mathrm{Var}(X),$$

since X and Y have the same distribution. Now this distribution is the simple one, with probability $\frac{1}{6}$ attached to each of the six values 1, 2, 3, 4, 5 and 6. Straightforward calculation then shows that $\mathrm{Var}(X) = \frac{35}{12}$, whence the required result.

The advantage of this method is that it extends easily to the sum of any number of dice, and not just two. The fact that for independent random variables variances can simply be added, as well as expectations, means that we can immediately write down, for W, the corresponding sum for n dice,

$$\mathrm{E}(W) = \frac{7n}{2},$$

and

$$\mathrm{Var}(W) = \frac{35n}{12}.$$

Note that in order to obtain these results we do not have to find the probability distribution of W; indeed this would be very difficult.

(2) The solution is presented in symbolic terms; we define various events and apply the laws of probability to them in order to obtain the probability distribution of Z. One could alternatively have presented the same method pictorially, representing the 36 outcomes of the experiment as a 6×6 grid of points. The sample space for the experiment will then consist of the 36 outcomes corresponding to these points, and to each point there is attached a probability ($\frac{1}{36}$ to each) and a value of Z (the sum of the values for X and Y). Summing the probabilities of outcomes with any particular value for Z will then give the probability of that value of Z, and hence the probability distribution.

1B.3 Selecting balls from an urn

An urn contains three red and five white balls. A ball is drawn at random, its colour is noted, and it is replaced along with another ball of the same colour. This process is repeated until three balls have been drawn. Find the mean and standard deviation of the number of red balls drawn.

Solution

We have to find the probability distribution of X, the number of red balls drawn. We begin by finding $\Pr(X = 0)$, which is equal to

$$\Pr(\text{first ball is white}) \times \Pr(\text{second ball is white} \mid \text{first ball is white})$$
$$\times \Pr(\text{third ball is white} \mid \text{first two balls are white})$$
$$= \frac{5}{8} \times \frac{6}{9} \times \frac{7}{10} = \frac{7}{24}.$$

Turning to $\Pr(X = 1)$, we find that there are three outcomes to consider. The first ball drawn may be red, and the second and third white; or the first may be white, followed by a red and another white; or two white balls may be followed by a red one. The first of these outcomes has probability

$$\Pr(\text{first ball is red}) \times \Pr(\text{second ball is white} \mid \text{first ball is red})$$
$$\times \Pr(\text{third ball is white} \mid \text{first is red, second is white})$$
$$= \frac{3}{8} \times \frac{5}{9} \times \frac{6}{10} = \frac{1}{8}.$$

We may similarly obtain the probabilities of the other two outcomes as

$$\frac{5}{8}\times\frac{3}{9}\times\frac{6}{10} = \frac{1}{8}$$

and

$$\frac{5}{8}\times\frac{6}{9}\times\frac{3}{10} = \frac{1}{8}$$

respectively. Summing the three probabilities we have obtained, we find that

$$\Pr(X = 1) = \frac{3}{8}.$$

The next value of X, 2, is dealt with similarly; again there are three outcomes to consider, depending on whether the single white ball drawn is in the first, second, or third position. We obtain

$$\begin{aligned}\Pr(X = 2) &= \left(\frac{5}{8}\times\frac{3}{9}\times\frac{4}{10}\right) + \left(\frac{3}{8}\times\frac{5}{9}\times\frac{4}{10}\right) + \left(\frac{3}{8}\times\frac{4}{9}\times\frac{5}{10}\right)\\ &= \frac{1}{4}.\end{aligned}$$

Finally,

$$\Pr(X = 3) = \frac{3}{8}\times\frac{4}{9}\times\frac{5}{10} = \frac{1}{12}.$$

Having obtained the probability distribution of X, it is now a simple matter to obtain its mean and standard deviation. Calculations are as follows:

$$\mathrm{E}(X) = \frac{(0\times7) + (1\times9) + (2\times6) + (3\times2)}{24} = \frac{27}{24} = \frac{9}{8};$$

$$\mathrm{E}(X^2) = \frac{(0\times7) + (1\times9) + (4\times6) + (9\times2)}{24} = \frac{51}{24} = \frac{17}{8}.$$

(The replacement, in the above, of the previously obtained values of the probability function of X by values expressed relative to a common denominator of 24 simplifies the computations somewhat.) Thus

$$\mathrm{Var}(X) = \frac{17}{8} - \left(\frac{9}{8}\right)^2 = \frac{55}{64},$$

and hence the standard deviation of X is

$$\sqrt{\frac{55}{64}} = 0{\cdot}927.$$

Notes

(1) The above solution, if written out in full without recourse to words like 'similarly', is rather lengthy. The use of set theoretic notation would enable us to produce a rather shorter solution. If we let R_i denote the event that the ith ball drawn is red, and W_i the event that it is white, we may write, for example,

$$\begin{aligned}\Pr(X = 2) = {} & \Pr(W_1)\times\Pr(R_2|W_1)\times\Pr(R_3|W_1\cap R_2)\\ & + \Pr(R_1)\times\Pr(W_2|R_1)\times\Pr(R_3|R_1\cap W_2)\\ & + \Pr(R_1)\times\Pr(R_2|R_1)\times\Pr(W_3|R_1\cap R_2).\end{aligned}$$

We may still decide to keep the length of the solution eventually presented within bounds by not writing out everything in detail, but the use of a notation such as this may help a lot when we are producing the detailed rough work that precedes a final solution. (See Note 1 to Problem 1A.2 and Note 5 to Problem 1A.10 for discussions of similar points.)

One might consider using a tree diagram to solve this problem. Since all branches of the tree would have to be explored in the course of solving the problem, there would not be the objection that there sometimes is on grounds of wasted effort. (See Note 6 to Problem 1A.10 for a discussion of the rôle of tree diagrams in the solution of problems in probability.)

(2) Notice that the three probabilities that are summed to obtain $\Pr(X = 1)$ are all the same. This suggests that we could get away with calculating just one component, and multiplying by three (on the grounds that the red ball could be drawn first, second, or third). The same applies to the case $X = 2$. An approach on these lines would, however, require justification; the dangers of assuming that it will work are illustrated by a variant of the present problem in which each ball is replaced along with one of the *other* colour: for this variant the approach of 'multiplying by three' will *not* work. (See also Note 2 to Problem 1A.2.)

1B.4 The game of darts

Three players A, B and C, take turns (in the specified order) to throw a dart at a dartboard; the first to hit the bull wins. In one throw, the probability that A hits the bull is $\frac{1}{10}$; the corresponding probabilities for B and C are $\frac{1}{9}$ and $\frac{1}{8}$ respectively.

(a) Show that the three players are equally likely to win.

(b) Find the expected number of throws in the game.

Solution

(a) The probability that A wins in the first round is clearly $\frac{1}{10}$. To obtain the probability that A wins in the second round we note that, if this is to happen, all three players must miss in the first round: the probability of this event is

$$(1 - \tfrac{1}{10})\times(1 - \tfrac{1}{9})\times(1 - \tfrac{1}{8}) = \tfrac{7}{10}.$$

Thus

$$\begin{aligned}\Pr(\text{A wins in round 2}) &= \Pr(\text{all three miss in round 1}) \\ &\quad \times \Pr(\text{A hits in round 2} \mid \text{all three miss in round 1}) \\ &= \tfrac{7}{10}\times\tfrac{1}{10}.\end{aligned}$$

Similarly

$$\begin{aligned}\Pr(\text{A wins in round 3}) &= \Pr(\text{all three miss in rounds 1 and 2}) \\ &\quad \times \Pr(\text{A hits in round 3} \mid \text{all three miss in rounds 1 and 2}) \\ &= (\tfrac{7}{10})^2\times\tfrac{1}{10}\end{aligned}$$

and, in general, for $i = 1, 2, \ldots,$

$$\Pr(\text{A wins in round } i) = \tfrac{1}{10}q^{i-1},$$

where we introduce the symbol q to represent the probability $\frac{7}{10}$ in the mathematical manipulations which follow. The probability that A wins is thus

$$\tfrac{1}{10}(1 + q + q^2 + \ldots) = \frac{\frac{1}{10}}{1 - q},$$

and substituting $\frac{7}{10}$ for q we find the probability to be $\frac{1}{3}$.

Turning our attention to B, we see that the probability that B wins in the first round is

$$\begin{aligned}\Pr(\text{A misses})\times\Pr(\text{B hits}) &= (1-\tfrac{1}{10})\times\tfrac{1}{9}\\ &= \tfrac{1}{10}.\end{aligned}$$

As with A, we continue to find that the probability that B wins in round i is $\frac{1}{10}q^{i-1}$; hence, summing over i, the probability that B wins is also $\frac{1}{3}$. Finally, by subtraction, the probability that C wins is $\frac{1}{3}$. (Alternatively, this result may be obtained using arguments similar to those above: the probability that C wins in round i is found to be the same as for the other two players.)

(b) From the results obtained in solving part (a), we see that the number of throws T in the game has the probability function

$$\Pr(T=i) = \begin{cases} \frac{1}{10}, & i = 1, 2, 3, \\ \frac{1}{10}q, & i = 4, 5, 6, \\ \frac{1}{10}q^2, & i = 7, 8, 9, \\ \text{etc.} \end{cases}$$

Hence

$$\begin{aligned}E(T) &= \tfrac{1}{10}\left(1+2+3+(4+5+6)q+(7+8+9)q^2+\ldots\right)\\ &= \tfrac{3}{10}(2+5q+8q^2+\ldots)\\ &= \tfrac{3}{10}(3+6q+9q^2+\ldots)\\ &\quad -\tfrac{3}{10}(1+q+q^2+\ldots)\\ &= \frac{\frac{9}{10}}{(1-q)^2}-\frac{\frac{3}{10}}{1-q}\\ &= \frac{\frac{9}{10}}{(1-\frac{7}{10})^2}-\frac{\frac{3}{10}}{1-\frac{7}{10}} = 10-1 = 9.\end{aligned}$$

Notes

(1) We are, throughout our solution, assuming that the results of different throws are independent.

(2) Some readers may wonder how we arrived at our solution to part (b). In finding the expectation and variance of a discrete probability distribution one frequently finds that the series to be summed are closely related to the series which one encounters in demonstrating that the probability function of the distribution sums to 1. Thus, for example, we encounter binomial series when dealing with the binomial distribution, exponential series when dealing with the Poisson distribution, and, in particular, binomial expansions of negative powers of $1-q$, where $q=1-p$, when dealing with the geometric distribution (see Problem 2A.10). Some links between the present distribution and the geometric distribution are explored briefly in Note 5.

To demonstrate that the probability function of T sums to 1, we note that the sum is

$$p(1+q+q^2+q^3+\ldots),$$

where $p=\frac{3}{10}=1-q$. Since the series in brackets can be recognised as the binomial expansion of $(1-q)^{-1}$ it is possible that, in seeking $E(T)$, we may find other binomial expansions of $1-q$ useful; in particular, we recollect that

$$(1-q)^{-2} = 1 + 2q + 3q^2 + 4q^3 + \ldots .$$

The series we require to evaluate in finding E(T) has coefficients (2, 5, 8, . . .) in arithmetic progression, with a common difference of 3, and the expansion of $3(1-q)^{-2}$ has the same property. Thus the difference between these series must be a series with all coefficients equal, i.e. a multiple of $(1-q)^{-1}$.

An alternative approach, which some may prefer, is the following. The series to be summed is

$$S = 2 + 5q + 8q^2 + 11q^3 + \ldots$$

(then $\mathrm{E}(T) = \frac{3}{10}S$). Noting that

$$qS = 2q + 5q^2 + 8q^3 + \ldots$$

and subtracting, we obtain

$$(1-q)S = 2 + 3q + 3q^2 + 3q^3 + \ldots .$$

We may now recognise that the right hand side of this equation is equal to $2 + 3q(1-q)^{-1}$, and hence proceed to evaluate S. Alternatively, we may repeat the process of multiplying by q and subtracting, obtaining

$$(1-q)^2 S = 2 + q.$$

Substituting the value $\frac{7}{10}$ for q, we obtain $S = 30$, and hence $\mathrm{E}(T) = 9$.

(3) There is another way of solving part (a). Since the argument is similar to that described in Note 2 to Problem 1A.6, we shall not give full details here. The basis of the method is that we condition on the fact that the game finishes in round k ($k = 1, 2, \ldots$). If the game has not ended before round k, the probabilities that A, B and C will win in this round are all $\frac{1}{10}$. Thus the conditional probabilities that they will win, given that the game ends in this round, are all equal to $\frac{1}{3}$; since these do not depend on k, they are also the unconditional probabilities.

(4) Another way of solving the problem is through the use of *recurrence relationships*. As our first illustration of this approach, we shall find the probability that A wins – call this p_{A}. Then

$$\begin{aligned} p_{\mathrm{A}} &= \Pr(\text{A wins in round 1}) + \Pr(\text{A wins in a subsequent round}) \\ &= \tfrac{1}{10} + \Pr(\text{all three miss in round 1}) \\ &\qquad \times \Pr(\text{A wins in a subsequent round} \mid \text{all three miss in round 1}) \\ &= \tfrac{1}{10} + \tfrac{7}{10}p_{\mathrm{A}}. \end{aligned}$$

We have now obtained a recurrence relationship for p_{A} which, on rearrangement, gives the equation $\frac{3}{10}p_{\mathrm{A}} = \frac{1}{10}$, with the solution $p_{\mathrm{A}} = \frac{1}{3}$.

A similar approach is also possible with part (b). If we let E denote the expected number of throws in the game, then

$$E = \left(\tfrac{1}{10}\times 1\right) + \left(\tfrac{1}{10}\times 2\right) + \left(\tfrac{1}{10}\times 3\right) + \left(\tfrac{7}{10}\times(3+E)\right).$$

(This follows from the fact that the number of throws in the game takes each of the values 1, 2 and 3 with probability $\frac{1}{3}$, and otherwise, with probability $\frac{7}{10}$, the game effectively 'restarts' after the first three throws.) The above recurrence relationship simplifies to $\frac{3}{10}E = \frac{27}{10}$, leading to $E = 9$.

(5) In solving part (b) we could have made use of the fact that the number of rounds in the game (including the final, possibly incomplete, round) has a geometric distribution with parameter $p = \frac{3}{10}$. The expected number of rounds in the game is therefore $\frac{10}{3}$. Now there will

be three throws in all but the last round; in the last the expected number of throws is 2. Thus

$$E(T) = (\tfrac{10}{3} - 1)\times 3 + 2 = 9.$$

Finally, we remark that the method of Note 4 may be used to find the mean of the geometric distribution, as an alternative to the more usual method involving the summation of series.

1B.5 The randomised response experiment

Let θ denote the probability that a randomly sampled individual in some population voted Conservative in the last General Election. In a particular type of 'randomised response' experiment, each of a random sample of individuals from this population responds 'True' or 'False' to one of the following two statements.

(a) I voted Conservative at the last General Election.

(b) I did not vote Conservative at the last General Election.

A randomising device is used to ensure that the probabilities of responding to (a) and (b) are p and $1-p$ respectively, where p is known and $0<p<1$. If λ is the probability that an individual responds 'True', write down an expression for λ in terms of p and θ.

For a group of mathematics teachers attending a statistics course, 24 out of 43 responded 'True' in an experiment in which p was fixed at 0·3. Use these figures to estimate θ.

Solution

In a rudimentary but obvious notation, we obtain

$$\begin{aligned}\Pr(\text{'True'}) &= \Pr(\text{'True'} \cap \text{question is (a)}) + \Pr(\text{'True'} \cap \text{question is (b)}) \\ &= p\Pr(\text{'True' to (a)}) + (1-p)\Pr(\text{'True' to (b)}), \\ \text{i.e. } \lambda &= p\theta + (1-p)(1-\theta) = (2p-1)\theta + 1 - p, \qquad (*)\end{aligned}$$

since the only people who respond 'True' to (a) are those who did vote Conservative, and similarly only those who did not vote Conservative respond 'True' to (b).

In the numerical example, we can estimate λ by

$$\hat{\lambda} = \frac{24}{43}.$$

Substituting this, and $p = 0{\cdot}3$, in equation (*) above and solving for θ gives $\hat{\theta} = 0{\cdot}355$ as an estimate of θ.

Notes

(1) This problem indicates a way of obtaining estimates of proportions of populations taking part in an 'embarrassing' activity; for example, exceeding speed limits regularly or smoking at school – outside the staff room, that is. We see from equation (*) that the method cannot work if p is set at 0·5. As long as p is not too close to 0 or 1 then the individual response of a subject does not give a strong indication of the true activity of that subject. With this restriction, the value of p is chosen so as to maximise the precision of the estimator of θ resulting from the experiment. In practice this means that p is chosen to be as close to 0 or 1 as is possible, while still making it clear to subjects that their true position will remain private.

The randomising device used could simply be a pack of cards, which the subject can take and shuffle as much as desired before selecting a card, or may be a bag of beads of different colours. The only important property of the randomising procedure is, of course, that it should be impossible (and clear to the subject that it is impossible) for the interviewer to know whether statement (a) or (b) was chosen by the subject.

(2) From Note 1, we see that it is important that the rules of the experiment be clear to the subjects. A preliminary experiment with the mathematics teachers mentioned in the problem produced (from the 51 then taking part in the experiment) an estimate $\hat{\lambda} = 35/51$, which gives $\hat{\theta} = 0{\cdot}034$. But in addition to performing the randomised response experiment, the participants also wrote down on a slip how they had voted, and this gave a direct estimate of θ of $21/51 = 0{\cdot}41$. One can use this latter figure to derive a confidence interval for θ, and a 95% interval (0·27, 0·55) was obtained by the first method shown in the solution to Problem 4C.3. We see that the value 0·034 is very far away from this, which, in a crude way, suggests that the technique might not have been adequately explained to the teachers. The value 0·355 obtained in a later experiment is, of course, quite consistent with the confidence interval above.

The two estimates of θ found above, 0·41 and 0·034, can be compared more formally, and correctly, as follows. For the 21 teachers who admitted to voting Conservative, the probability of responding 'True' in the randomised response experiment is 0·3; for the other 30 teachers, the probability of responding 'True' is 0·7. So if X is a random variable denoting the total number responding 'True', we can write

$$X = X_C + X_{\bar{C}}$$

where X_C and $X_{\bar{C}}$ are independent binomially distributed random variables with distributions $B(21, 0{\cdot}3)$ and $B(30, 0{\cdot}7)$ respectively. Hence

$$\mathrm{E}(X) = (21\times0{\cdot}3) + (30\times0{\cdot}7) = 27{\cdot}3$$

$$\text{and} \quad \mathrm{Var}(X) = (21\times0{\cdot}3\times0{\cdot}7) + (30\times0{\cdot}7\times0{\cdot}3) = 10{\cdot}71.$$

Using a normal approximation to the distribution of X, we find that, approximately, $X \sim N(27{\cdot}3, 10{\cdot}71)$. The observed value of X from the randomised response experiment was 35, and the corresponding standardised normal deviate z is

$$z = \frac{35 - \frac{1}{2} - 27{\cdot}3}{\sqrt{10{\cdot}71}} = 2{\cdot}20,$$

using a continuity correction since X is discrete. This value of z therefore shows a significant difference at the 5% level but not at the 1% level.

(3) There are many variations on the basic randomised response technique. See, for example, Problem 4D.6 for a different version, in which statement (b) above is replaced by an 'innocent' statement, concerning the timing of the subject's birthday. As will be seen in the solution to Problem 4D.6, the procedure is very much as in the present problem, but one needs to know the probability of replying 'True' to the innocent statement. (If this probability is unknown, one can undertake the randomised response experiment twice, using different values of p, which will give two equations to be solved for the two unknown probabilities.)

1B.6 Finding a probability density function

A continuous random variable X, with mean unity, has probability density function $f_X(x)$ given by

$$f_X(x) = \begin{cases} a(b-x)^2, & 0 \le x \le b, \\ 0, & \text{otherwise.} \end{cases}$$

Find the values of a and b.

Solution

In order to solve for the two unknown values a and b, we need two equations involving a and b. These equations are obtained from the two items of information that we are given, as follows.

(i) The function $f_X(x)$ is a probability density function. Hence

$$\int_0^b a(b-x)^2 \mathrm{d}x = 1,$$

and so $$\left[\frac{-a(b-x)^3}{3}\right]_0^b = 1,$$

resulting in the equation $ab^3 = 3$.

(ii) The mean of X is unity. So

$$\int_0^b a(b-x)^2 x\,\mathrm{d}x = 1,$$

and thus

$$a\int_0^b (b-x)^2(x-b+b)\mathrm{d}x = 1.$$

Splitting terms in the second bracket gives

$$-a\int_0^b (b-x)^3 \mathrm{d}x + ab\int_0^b (b-x)^2 \mathrm{d}x = 1,$$

and these integrals are evaluated as above to yield

$$-\frac{ab^4}{4} + \frac{ab^4}{3} = 1,$$

which simplifies to give $ab^4 = 12$. This is to be solved together with the equation $ab^3 = 3$ obtained earlier. We now see that $b = 4$ and $a = \frac{3}{64}$.

Note

In some respects this problem might be thought rather unrealistic, in that one rarely finds a random variable with a quadratic density function in nature. (Such random variables are, however, of value in simulation, where they are used as a basis for generating other distributions, and in particular the normal distribution.) If a histogram of observations on some random variable showed symmetry, and the range was restricted, one might perhaps consider using a quadratic density function as a rough approximation.

The most direct value in this problem resides mainly in the reinforcement it can give, using only elementary calculus, to the result that all probability density functions integrate to 1 over the relevant range, and to the formula $\int x\, f_X(x)\mathrm{d}x$ for the expectation of a random variable X.

1B.7 Positioning the pointer

In a psychological experiment investigating how individuals change their assessments of probability in the light of data, subjects have to indicate their probabilities for certain events by moving a pointer on a scale between 0 and 1. Consider a subject who is unable to give any worthwhile assessment of a particular probability, and decides to place the pointer randomly, with each point in $(0, 1)$ equally likely to be the point chosen. In such a case, what is the probability that the ratio of the resulting shorter segment to the longer segment is less than $\frac{1}{4}$?

Solution

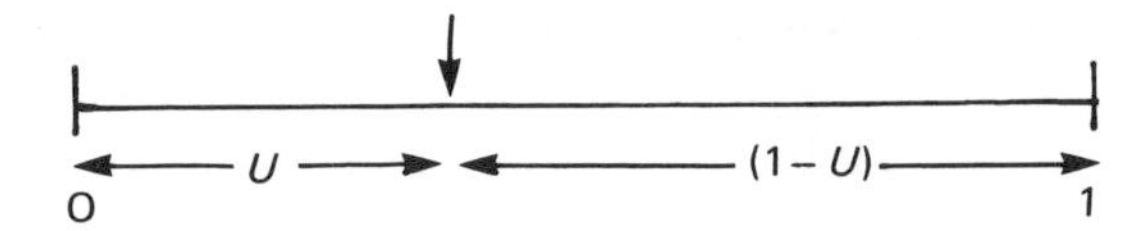

Figure 1.4 Position of pointer on (0, 1) scale

Let U denote the distance from 0 to the pointer (see Figure 1.4). From the symmetry of the problem it is sufficient to consider the case $U < \frac{1}{2}$; we thus require

$$\begin{aligned}\Pr\left(\frac{U}{1-U} < \tfrac{1}{4} \mid U < \tfrac{1}{2}\right) &= \Pr(4U < 1 - U \mid U < \tfrac{1}{2}) \\ &= \Pr(U < \tfrac{1}{5} \mid U < \tfrac{1}{2}) \\ &= \frac{\Pr(U < \frac{1}{5})}{\Pr(U < \frac{1}{2})} \\ &= \tfrac{1}{5}/\tfrac{1}{2} = \tfrac{2}{5}.\end{aligned}$$

Notes

(1) An alternative way of obtaining the solution is as follows. The event of interest occurs when $U < \frac{1}{5}$ or, by symmetry, when $U > \frac{4}{5}$. The required probability is therefore

$$\Pr(U < \tfrac{1}{5}) + \Pr(U > \tfrac{4}{5}) = \tfrac{2}{5}.$$

(2) Experiments on the assessment and revision of probabilities are easily carried out and can be revealing. Suppose a bag of 10 balls is used, 7 of the balls being red and 3 black. Before drawing any balls from the bag, a subject's estimate of the probability that a ball drawn at random is red may be 0·5. The subject may then successively select balls at random, observe their colour and replace them in the bag, each time revising his or her estimate of the probability of drawing a red ball. Experiments of this kind have shown that subjects behave *conservatively* in comparison with the predictions of Bayes' Theorem.

* (3) A random variable with a uniform distribution, conditioned as in this problem, remains a uniform random variable.

This result is sometimes useful in the area of simulation. For example, suppose we want to simulate a random variable X with the binomial distribution $B(2, \frac{1}{2})$. One way of doing this would be to select two independent random variables, U_1 and U_2, say, both uniformly distributed over the range $(0, 1)$. We then set $X = I_1 + I_2$, where $I_i = 1$ if $U_i > \frac{1}{2}$ and 0 otherwise, for $i = 1, 2$. For each value of X we need to generate two uniform random variables, and quite often generating such variables is a time-consuming feature of simulation.

We now see that we do not need to select a new value for U_2, but rather we can *re-use* U_1. Thus if, say, $U_1 < \frac{1}{2}$, then $2U_1$ is uniformly distributed over $(0, 1)$, and may thus be used in place of U_2.

We can verify this as follows. For $0 \le x \le 1$,

$$\Pr(2U_1 < x \mid U_1 < 0{\cdot}5) = \frac{\Pr(2U_1 < x \cap U_1 < 0{\cdot}5)}{\Pr(U_1 < 0{\cdot}5)}$$

$$= \frac{\Pr(U_1 < \frac{1}{2}x)\Pr(U_1 < 0\cdot5 \mid U_1 < \frac{1}{2}x)}{\Pr(U_1 < 0\cdot5)}.$$

Now, since $0 \le x \le 1$, $0 \le \frac{1}{2}x \le \frac{1}{2}$, and so $U_1 \le \frac{1}{2}x$ implies that $U_1 \le \frac{1}{2}$. So we obtain

$$\Pr(2U_1 < x \mid U_1 < 0\cdot5) = \frac{\frac{1}{2}x \times 1}{\frac{1}{2}} = x, \text{ as required.}$$

As an illustration, suppose that $U_1 = 0\cdot8442$. Since $U_1 > 0\cdot5$ we set $I_1 = 1$. In this case $U_1 > 0\cdot5$, but $1 - U_1 < 0\cdot5$, and so in place of U_2 we can take $2(1 - U_1) = 0\cdot3116 < 0\cdot5$. We therefore set $I_2 = 0$, and so obtain the simulated value of $X = I_1 + I_2 = 1$.

1B.8 Finding the mode and median of a distribution

A continuous random variable X has probability density function $f_X(x)$ given by

$$f_X(x) = k(2 - x)(x - 5), \qquad 2 \le x \le 5,$$
$$= 0, \qquad \text{elsewhere.}$$

Find the value of k, and hence deduce the mean and variance of X. What are the values of the mode and median of the distribution of X?

Solution

Since $f_X(x)$ is a probability density function, the area under it must be unity, i.e.

$$\int f_X(x)\,dx = 1,$$

where the integral is taken over the range of the random variable X. We can clearly use this result to evaluate k, as follows.

$$\int_2^5 k(2 - x)(x - 5)\,dx = 1,$$

$$\text{i.e.}\quad k\int_2^5 (-x^2 + 7x - 10)dx = 1.$$

$$\text{Then}\quad k\left[-\frac{x^3}{3} + \frac{7x^2}{2} - 10x\right]_2^5 = 1,$$

and this gives $k \times \frac{9}{2} = 1$ or $k = \frac{2}{9}$.

The mean of X is defined as

$$E(X) = \int_2^5 x\,f_X(x)\,dx = \frac{2}{9}\int_2^5 x(2 - x)(x - 5)dx$$

$$= \frac{2}{9}\left[-\frac{x^4}{4} + \frac{7x^3}{3} - 5x^2\right]_2^5 = \frac{7}{2}.$$

The variance of X may be written as $\mathrm{Var}(X) = E(X^2) - \{E(X)\}^2$. Thus we require

$$E(X^2) = \int_2^5 x^2 f_X(x)dx = \frac{2}{9}\int_2^5 x^2(2 - x)(x - 5)\,dx$$

$$= \frac{2}{9}\left[-\frac{x^5}{5} + \frac{7x^4}{4} - \frac{10x^3}{3}\right]_2^5 = \frac{127}{10},$$

leading to

$$\text{Var}(X) = \frac{127}{10} - \frac{49}{4} = \frac{9}{20}.$$

The density function in this problem is parabolic, and so we have a single mode, m, given by the value of x which maximises $f_X(x)$ over the range $2 \le x \le 5$. Now

$$\frac{\mathrm{d}f_X(x)}{\mathrm{d}x} = k(-2x+7)$$

and so $\dfrac{\mathrm{d}f_X(x)}{\mathrm{d}x} = 0$ when $x = \frac{7}{2}$. Thus the mode of the distribution is at $x = \frac{7}{2}$. (Differentiating again shows that $\dfrac{\mathrm{d}^2 f_X(x)}{\mathrm{d}x^2} = -7k$. This is always negative, and in particular when $x = \frac{7}{2}$, providing confirmation that the stationary value is a maximum.)

In order to determine the median, M, we need to solve the equation

$$F_X(M) = \int_2^M f_X(x)\,\mathrm{d}x = \frac{1}{2},$$

from the definition of the median. Thus

$$\frac{2}{9}\int_2^M (2-x)(x-5)\,\mathrm{d}x = \frac{1}{2},$$

so that

$$\frac{2}{9}\left[-\frac{x^3}{3} + \frac{7x^2}{2} - 10x\right]_2^M = \frac{1}{2},$$

leading to

$$\frac{2}{9}\left(-\frac{M^3}{3} + \frac{7M^2}{2} - 10M + \frac{26}{3}\right) = \frac{1}{2}.$$

This equation reduces to the cubic

$$4M^3 - 42M^2 + 120M - 77 = 0. \qquad (*)$$

The equation has solution $M = \frac{7}{2}$ (see Note 2) and so we may factorise the cubic as

$$(2M-7)(2M^2 - 14M + 11) = 0.$$

The equation $2M^2 - 14M + 11 = 0$ has roots 0·902 and 6·098, both of which are unacceptable since we must have $2 \le M \le 5$. The median is therefore given by $M = \frac{7}{2}$.

Notes

(1) This problem is useful in indicating the type of manipulation needed to calculate the population mean, variance, mode and median for a simple probability density function.

(2) In this example the mean, mode and median all coincide at $x = \frac{7}{2}$. The reason for this result is simply that the probability density function $f_X(x)$, being parabolic, is symmetrical, as shown in Figure 1.5. Had this graph been drawn initially then the common identity of the mean, mode and median at $x = \frac{7}{2}$ would have been spotted instantly. Indeed, had this fact not been known then the root $M = \frac{7}{2}$ to the cubic equation (*) would not have been obvious. We see therefore the value of a rough graph, in shedding light on otherwise purely algebraic manipulations which might well have produced the wrong answer from an arithmetic error.

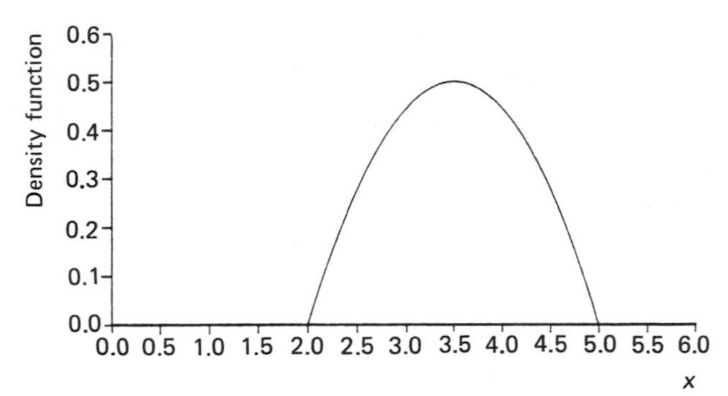

Figure 1.5 Probability density function for Problem 1B.8

(3) As $F_X(x)$ is a continuous, increasing, function of x in the range $2 \le x \le 5$, the equation $F_X(x) = \frac{1}{2}$ can only have one root for x in this range. Once the root $x = \frac{7}{2}$ is known then, strictly, no further analysis is necessary to find the median.

1B.9 Buffon's needle

Two infinitely long parallel lines are a distance a apart, and a needle of length l ($l < a$) is dropped and spun so as to fall with its midpoint equally likely to lie at any point between the lines, and equally likely to point in any direction. Show that the probability that the needle crosses one of the lines is $\dfrac{2l}{\pi a}$.

Solution

This complex problem involves two jointly distributed random variables: D, the distance of the midpoint from one of the lines, and Θ, the angle made by the needle with the lines. From the conditions stated, D is uniformly distributed between 0 and a while, independent of D, Θ is uniformly distributed between 0 and π.

The simplest solution follows from a conditional argument. Suppose that, on a particular throw of the needle, Θ takes the value θ. Then, given this, the needle will cross a line if its midpoint lies closer than $\frac{l}{2}\sin\theta$ to either of the two lines, i.e.

$$\Pr(A \mid \Theta = \theta) = \frac{\frac{l}{2}\sin\theta + \frac{l}{2}\sin\theta}{a},$$

where A is the event that the needle crosses a line. Hence

$$\Pr(A) = \int_0^\pi \Pr(A \mid \Theta = \theta) f_\Theta(\theta)\, d\theta \qquad (*)$$

$$= \int_0^\pi \frac{l\sin\theta}{a} \frac{1}{\pi}\, d\theta$$

$$= \frac{l}{\pi a}\int_0^\pi \sin\theta\, d\theta = \frac{2l}{\pi a}.$$

Notes

(1) The argument in equation (*) is a form (strictly, an extension) of the law of total probability, also discussed in Problems 1A.4 and 1A.7, amongst others. This probability could have been arrived at by examining the joint probability density function of D and Θ and integrating it over the relevant region. But the double integral involved is avoided by the conditional argument; indeed it often simplifies an analysis to look for a random variable on which one can condition.

(2) This problem, in which a needle is dropped randomly onto a grid (simplified here to just two lines, without loss of generality), is given the name of Buffon's needle, and the corresponding experiment can be used to estimate π. Suppose that a needle is thrown randomly n times onto the grid; let X be the number of occasions on which a line is crossed. Then (under reasonable assumptions) X will be binomially distributed with index n and parameter (i.e. probability of success) $\dfrac{2l}{\pi a}$. The fraction $\dfrac{X}{n}$ will then give an estimate of this probability, and one can transform this in a simple way to give an estimate $\dfrac{2ln}{aX}$ for π. (There is an obvious problem if no 'successes' occur; we do not deal with this here, beyond commenting that by choosing l, a and n appropriately such an eventuality can be made extremely unlikely.)

(3) The experiment can be extended and modified in various ways, of which we mention just a few.

(i) *The double grid.* Suppose that the needle falls randomly onto a grid of squares of size $a \times a$. Clearly there are now (for a needle of length $l \leq a$) three possibilities: that the needle falls entirely within a square (probability $1 - (4r - r^2)/\pi$), where $r = l/a$, that it crosses exactly one line ($2(2r - r^2)/\pi$), and two lines (r^2/π). Results from an experiment can then be used in a variety of ways to produce an estimate of π. For example, one might equate the proportion crossing two lines to r^2/π, or the proportion crossing at least one line (i.e. not lying entirely within a square) to $(4r - r^2)/\pi$.

The experiment thus offers many opportunities for class discussion. For example, different members of the group may try the experiment with different values of r, and for any set of results both the estimates noted above might be calculated. Comparison of results will then throw light on the properties of the various estimators, and this in turn can give a basis for a discussion of concepts such as efficiency of estimators.

(ii) *Needles of length greater than a.* When the length l of the needle exceeds the distance a between lines the situation is naturally more complicated. But in one respect it remains very simple. If in the original solution we let X denote the number of lines crossed by the needle, then plainly X can be only 0 or 1, and

$$\mathrm{E}(X) = 0 \times \Pr(X = 0) + 1 \times \Pr(X = 1) = \Pr(A) = \frac{2l}{\pi a}.$$

Now suppose a long needle has bands painted across it so that the lengths $l_1, l_2, \ldots, l_m$ of the m individual sections are all less than a. If we denote by X_i the number of lines (0 or 1) crossed by the section of length l_i, and if the entire needle crosses X lines, then

$$X = X_1 + X_2 + \ldots + X_m,$$

and so

$$\begin{aligned}\mathrm{E}(X) &= \mathrm{E}(X_1) + \mathrm{E}(X_2) + \ldots + \mathrm{E}(X_m)\\ &= \frac{2}{\pi a}(l_1 + l_2 + \ldots + l_m) = \frac{2l}{\pi a}.\end{aligned}$$

In other words the same value is calculated now as before, but for a long needle it represents the *mean* number of lines crossed, rather than the *probability* of a line being crossed.

(iii) *Curved needles.* In our discussion of long needles above we made no use of the customary property of needles of being straight. In principle, each individual section of the needle has to be straight, but there is no requirement that the whole needle shares this property. Further, we are quite entitled to have the individual sections of infinitesimal length, so that by a limiting argument the needle can be of any (two-dimensional!) shape.

An interesting special case, which we leave to the reader to develop, is that in which the needle is in the shape of a circle.

(4) The basic single and double grid experiments have been used by the authors in courses at the University of Kent over several years. It must be said that the results have not always been regarded as convincing proof that the value of π can safely be left in the care of statisticians, although an unscrupulous operator could select results so as to show that good estimates can be produced. For example, in the single grid experiment, results over three years' courses for teachers gave 2487 lines crossed out of 3890 trials in which $l = a$, i.e. the needle was the same length as the distance between lines; this gives an estimate of the probability that a line is crossed to be 0·639 and a corresponding estimate of π as 3·128. In the double grid experiment, over two years, there were 3110 trials, of which 153 resulted in the needle lying entirely within the square, 1032 had both lines crossed, and in the remaining 1933 only one of the lines was crossed. The estimate of π based on the number crossing both lines is $\frac{3\times 3110}{3110-153}$, which gives 3·155.

Bearing in mind the link with the binomial distribution, we see that the results above can be manipulated to give tests of hypotheses (or, more likely, confidence intervals) for the probabilities of the various relevant events, along the lines shown in Problems 4C.2 and 4C.3. Since these probabilities are all straightforward functions of π, corresponding inferences can be drawn about π. This can again lead on to discussions of efficiency of estimation, optimum choice of r, and so on.

1B.10 Transforming a random variable

Find the mean and variance of the continuous random variable X with probability density function given by

$$f_X(x) = \begin{cases} 3x^{-4}, & x \geq 1, \\ 0, & \text{otherwise.} \end{cases}$$

A new random variable Y is defined by the relation $Y = X^{-1}$. Find the cumulative distribution function of Y, and hence derive its probability density function.

Verify that, for X and Y as given above, $\mathrm{E}(XY) \neq \mathrm{E}(X)\,\mathrm{E}(Y)$, and explain why this result holds in this particular example.

Solution

By definition,

$$\mathrm{E}(X) = \int_{-\infty}^{\infty} x f_X(x)\mathrm{d}x = 3\int_{1}^{\infty} x^{-3}\mathrm{d}x = \frac{3}{2},$$

and

$$\begin{aligned}\mathrm{Var}(X) &= \mathrm{E}(X^2) - \{\mathrm{E}(X)\}^2 \\ &= 3\int_{1}^{\infty} x^{-2}\mathrm{d}x - \frac{9}{4} = 3 - \frac{9}{4} = \frac{3}{4}.\end{aligned}$$

The cumulative distribution function of Y is given by $F_Y(y) = \Pr(Y \leq y)$. We can see that the two events $\{Y \leq y\}$ and $\{X \geq y^{-1}\}$ are identical, and so, for $0 \leq y \leq 1$,

$$\begin{aligned}\Pr(Y \leq y) &= \Pr(X \geq y^{-1}) \\ &= 3\int_{y^{-1}}^{\infty} x^{-4}\mathrm{d}x \\ &= y^3.\end{aligned}$$

The probability density function of Y is then given by

$$f_Y(y) = \frac{\mathrm{d}}{\mathrm{d}y}\Pr(Y \leq y) = \begin{cases} 3y^2, & 0 \leq y \leq 1, \\ 0, & \text{otherwise.} \end{cases}$$

We have already found $\mathrm{E}(X)$ to be $\frac{3}{2}$, and now we obtain

$$\mathrm{E}(Y) = \int_0^1 y\, 3y^2\mathrm{d}y = \frac{3}{4},$$

so, clearly, $\mathrm{E}(X)\,\mathrm{E}(Y) = \frac{9}{8}$. But, by definition, $XY = 1$ always, so

$$\mathrm{E}(XY) = 1 \neq \tfrac{9}{8} = \mathrm{E}(X)\,\mathrm{E}(Y).$$

Had we found that $\mathrm{E}(XY) = \mathrm{E}(X)\,\mathrm{E}(Y)$, then the covariance between X and Y would have been zero, as would the correlation. However, Y is a monotonically decreasing function of X, so that X and Y are clearly correlated – negatively, as one would expect.

Notes

(1) When dealing with unusual density functions, and also when deriving new ones, it is often useful to check that (as required for a density function) they integrate to unity. Here

$$\int_{-\infty}^{\infty} f_X(x)\mathrm{d}x = 3\int_1^{\infty} x^{-4}\mathrm{d}x = 1, \quad \text{and}$$

$$\int_0^1 f_Y(y)\mathrm{d}y = 3\int_0^1 y^2\mathrm{d}y = 1.$$

(2) It is useful to remember that the probability density function of a new random variable is often readily obtained by first deriving its cumulative distribution function and then differentiating, as was done here. Another example is given in Problem 1B.11.

(3) Notwithstanding the comment in Note 2, there is another, quicker, method of deriving the density function of a random variable defined (as in this problem) through a transformation. The method is available only when the function $Y = g(X)$ defining the transformation is strictly monotonic, i.e. its derivative always takes the same sign. When this happens it can be shown that, given the density function $f_X(x)$ of X and the transformation $y = g(x)$,

$$f_Y(y) = \frac{f_X(x)}{\left|\frac{\mathrm{d}y}{\mathrm{d}x}\right|}, \qquad (*)$$

where we express the right hand side in terms of y by solving the equation $y = g(x)$ for x and substituting for x.

For the current problem, $f_X(x) = 3x^{-4}$ and the transformation is $y = x^{-1}$, so that $\frac{\mathrm{d}y}{\mathrm{d}x} = -x^{-2}$. We thus obtain, from equation (*) above,

$$f_Y(y) = \frac{3x^{-4}}{|-x^{-2}|} = 3x^{-2},$$

and substituting y^{-1} for x gives $f_Y(y) = 3y^2$, as before.

(4) In problems of this type it is almost always far simpler to use the expression

$$\mathrm{Var}(X) = \mathrm{E}(X^2) - \{\mathrm{E}(X)\}^2$$

when calculating a variance, rather than the alternative and equivalent form

$$\mathrm{Var}(X) = \mathrm{E}[\{X - \mathrm{E}(X)\}^2].$$

An exception is discussed in Note 1 to Problem 1B.1; some further comments on calculating variances are made in Note 1 to Problem 2A.2.

1B.11 Length of a chord of a circle

In Figure 1.6, OP is a radius of a circle with centre O and radius r. A point Q is chosen on OP in such a way that all points on OP have the same chance of being chosen. The chord passing through Q at right angles to OP meets the circle at points A and B. Find the probability density function of the length AB, and show that the probability that AB is greater than r is 0·866.

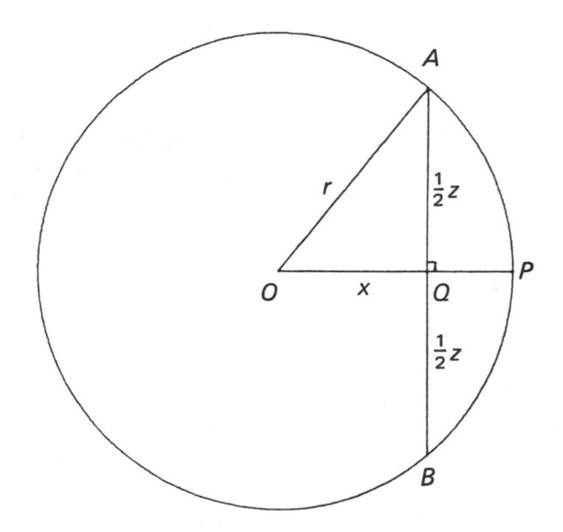

Figure 1.6

Solution

Since the point Q is randomly positioned, the distances OQ and AB are both random variables, and we denote them by X and Z respectively. We are given that X is uniformly distributed on the range $(0, r)$, and require the distribution of Z. The random variable Z is, of course, a function of X; by Pythagoras' Theorem we see that

$$X^2 + (\tfrac{1}{2}Z)^2 = r^2.$$

In problems of this sort the simplest approach is to obtain first the distribution function $F_Z(z)$ of Z, i.e. $\Pr(Z \leq z)$. Now, from the above equation, $Z = 2\sqrt{r^2 - X^2}$, so the event $\{Z \leq z\}$ can be written as $\{2\sqrt{r^2 - X^2} \leq z\}$, or as $\{X \geq \sqrt{r^2 - z^2/4}\}$. Since these events are identical, they must have the same probability, so

$$F_Z(z) = \Pr(Z \leq z) = \Pr(X \geq \sqrt{r^2 - z^2/4}).$$

We now use the fact that X is equally likely to take any value in the range $(0, r)$. The probability that it lies between $\sqrt{r^2 - z^2/4}$ and r is thus

$$\frac{r - \sqrt{r^2 - z^2/4}}{r},$$

so that

$$F_Z(z) = 1 - \frac{\sqrt{r^2 - z^2/4}}{r}.$$

To obtain the probability density function of Z we merely differentiate. Hence

$$f_Z(z) = \frac{\mathrm{d}}{\mathrm{d}z} F_Z(z) = -\frac{1}{2r}\left(r^2 - \frac{z^2}{4}\right)^{-\frac{1}{2}}\left(-\frac{2z}{4}\right)$$

$$= \frac{z}{4r}\left(r^2 - \frac{z^2}{4}\right)^{-\frac{1}{2}}, \qquad 0 \le z < 2r.$$

To evaluate $\Pr(Z > r)$ it is simplest to write

$$\Pr(Z > r) = 1 - \Pr(Z \le r)$$

$$= 1 - \left\{1 - \frac{\sqrt{r^2 - r^2/4}}{r}\right\} = \frac{\sqrt{r^2 - r^2/4}}{r} = \sqrt{0{\cdot}75} = 0{\cdot}866,$$

as required.

Notes

(1) Like Problem 1B.10, this problem deals with the transformation of random variables. Some questions of this type are readily answered by working directly with the probability density function, as is demonstrated in Note 3 to Problem 1B.10. In other cases, as here, it is simpler to work with the cumulative distribution function.

This problem is not an easy one; in particular, attempts to solve it by working with trigonometric functions of the angle *AOP* could result in difficulties.

(2) The answer to the second part of the problem could have been obtained by integrating the density function of Z over the range from r to $2r$, but this was not necessary here.

(3) In the solution we used an intuitive method to evaluate $\Pr(X \ge \sqrt{r^2 - z^2/4})$. There is, of course, a rigorous approach available, which some readers may prefer. Since X is uniformly distributed on the range $(0, r)$, its density function is given by

$$f_X(x) = \frac{1}{r}, \qquad 0 \le x \le r,$$

and is 0 elsewhere. Thus X has cumulative distribution function

$$F_X(x) = \frac{x}{r}, \qquad 0 \le x \le r.$$

But

$$\Pr(X \ge \sqrt{r^2 - z^2/4}) = F_X(r) - F_X(\sqrt{r^2 - z^2/4})$$

$$= \frac{r}{r} - \frac{\sqrt{r^2 - z^2/4}}{r} = \frac{r - \sqrt{r^2 - z^2/4}}{r},$$

as found in the solution.

2 Probability Distributions

In this chapter we make the short jump from the general examination of random variables and their distributions to dealing with some of the most important probability distributions; the chapter will concentrate particularly on the binomial, Poisson and normal distributions. Naturally we also give some discussion to others, if only to encourage readers to avoid habits of thought which may dictate, for example, that if the binomial distribution cannot be used in a problem involving a discrete random variable, then the distribution must be Poisson, which is certainly not the case.

A feature of the problems in this chapter is that many involve two or more distributions. Such problems are by no means artificial. Indeed, as we shall see, one of the skills of the specialist in probability theory is to appreciate the opportunities to save effort in complex problems through breaking them down into manageable components, each involving, perhaps, just a single distribution. A closely related skill, also involved in some of these problems, is that of appreciating the relationships, both exact and approximate, between different distributions.

2A Discrete Distributions

For discrete random variables, particularly, the art of getting to grips with a problem is largely that of recognising circumstances leading to each of the well-known probability distributions. So, for example, if in a problem one can identify independent 'trials', each with the same probability of a 'success', then the total number of these 'successes' in a fixed number of 'trials' (whatever these represent in the context of the problem) has a binomial distribution. This section concentrates on the binomial and Poisson distributions, when it is appropriate to use them, and when it is suitable to use an approximation. In some cases we use one discrete distribution to approximate another, if that is appropriate and seems useful. In other cases an approximation using the normal distribution is used.

It is convenient to deal here with a few small points of notation for these distributions. We shall use the notation '$X \sim B(n,p)$' to denote that a random variable X has the binomial distribution with index n and parameter p; the *index* of a binomial distribution is the number of 'trials', and the *parameter* is the probability of 'success'. Similarly, if a random variable Y has a Poisson distribution with mean μ we shall write $Y \sim \text{Poisson}(\mu)$. (When using a normal distribution approximation we shall also, naturally, make use of appropriate notation; for convenience, we defer a full description of normal distribution notation until the introduction to Section 2B, but note briefly here that we use the virtually standard notation $N(\mu, \sigma^2)$ for the distribution and $\Phi(z)$ for the cumulative distribution function of the standardised normal distribution $N(0, 1)$.)

There are many other discrete distributions, and we also deal with some of these, for example the negative binomial, geometric and hypergeometric distributions.

2A.1 The squash match

Paul and Eric are playing squash, and Paul is determined to win at least two games. Unfortunately his chance of winning any one game is only $\frac{1}{4}$, and this chance remains constant however many games he plays against Eric. The players agree to play 5 games and, if Paul has won at least two by then, play ceases. Otherwise Paul persuades Eric to play a further 5 games with him. What is the probability

(i) that only 5 games are played, and Paul wins at least two of them;

(ii) that 10 games have to be played, and Paul wins at least two?

Solution

We denote the number of games Paul wins out of the first five games by X_1, and the number out of the second five (if played) by X_2. Then $X_1 \sim B(5, \frac{1}{4})$, and X_2 will have the same distribution.

(i) We require $\Pr(X_1 \geq 2)$, and this is most simply evaluated as

$$\begin{aligned}\Pr(X_1 \geq 2) &= 1 - \Pr(X_1 = 0) - \Pr(X_1 = 1)\\ &= 1 - (\tfrac{3}{4})^5 - 5(\tfrac{1}{4})(\tfrac{3}{4})^4\\ &= \frac{1024 - 243 - 405}{1024} = \frac{376}{1024} = 0{\cdot}367.\end{aligned}$$

(ii) We now require the probability that ten games are played, and that Paul wins at least two; we denote this event by S. From the conditions stated, X_1 must be 0 or 1, and $X_1 + X_2$ must be two or more. We thus obtain

$$\Pr(S) = \Pr(X_1 = 0 \cap X_2 \geq 2) + \Pr(X_1 = 1 \cap X_2 \geq 1),$$

by the addition law of probability for mutually exclusive events. Now each of these events is itself the intersection of two independent events, to which the multiplication law can be applied. Hence

$$\begin{aligned}\Pr(S) &= \left[(\tfrac{3}{4})^5 \times \frac{376}{1024}\right] + \left[5(\tfrac{1}{4})(\tfrac{3}{4})^4 \times \{1 - (\tfrac{3}{4})^5\}\right]\\ &= \left(\frac{243}{1024} \times \frac{376}{1024}\right) + \left(\frac{405}{1024} \times \frac{781}{1024}\right) = 0{\cdot}3888.\end{aligned}$$

Note

The binomial distribution could be used in the solution because it was clear that the first five games could be regarded as five independent 'trials', in each of which there were just two possible results, 'success' for Paul, and 'failure', with the same probability of success on each trial. (Trials of this type are often termed 'Bernoulli trials'.) Under these circumstances the total number of successes, X_1, is known to have a binomial distribution.

2A.2 Sampling incoming batches

Show that the binomial distribution with index n and parameter p has mean np and variance $np(1-p)$.

A company taking delivery of a large batch of manufactured articles accepts the batch if either (a) a random sample of 6 articles from the batch contains not more than one defective article, or (b) a random sample of 6 contains two defective articles, and a second random sample of 6 is taken, and found to contain no defectives. If 20% of the articles in the batch are actually defective, what is the probability that the company will accept the delivered batch?

Solution

If $X \sim B(n,p)$, then

$$\Pr(X=x) = \binom{n}{x} p^x (1-p)^{n-x}, \quad x = 0, 1, \ldots, n.$$

From the definition of expectation, we thus obtain

$$\mathrm{E}(X) = \sum_{x=0}^{n} x \binom{n}{x} p^x (1-p)^{n-x}.$$

But $x\binom{n}{x} = n\binom{n-1}{x-1}$, for $x = 1, 2, \ldots, n$, and the initial term $(x=0)$ in the summation is zero. Hence

$$\begin{aligned}\mathrm{E}(X) &= n \sum_{x=1}^{n} \binom{n-1}{x-1} p^x (1-p)^{n-x} \\ &= np \sum_{x=1}^{n} \binom{n-1}{x-1} p^{x-1} (1-p)^{(n-1)-(x-1)} \\ &= np,\end{aligned}$$

since substituting $y = x-1$ in the summation shows it to be just the binomial expansion of $\{p + (1-p)\}^{n-1}$, or, equivalently, the sum of probabilities in the $B(n-1,p)$ distribution.

To obtain $\mathrm{Var}(X)$, we first use an extension of the method above to obtain $\mathrm{E}\{X(X-1)\}$. We find

$$\begin{aligned}\mathrm{E}\{X(X-1)\} &= \sum_{x=0}^{n} x(x-1) \binom{n}{x} p^x (1-p)^{n-x}. \\ &= n(n-1) \sum_{x=2}^{n} \binom{n-2}{x-2} p^x (1-p)^{n-x} \\ &= n(n-1)p^2 \sum_{x=2}^{n} \binom{n-2}{x-2} p^{x-2} (1-p)^{(n-2)-(x-2)} \\ &= n(n-1)p^2,\end{aligned}$$

since using the substitution $y = x-2$ shows that the summation is 1. We can now deduce an expression for $\mathrm{Var}(X)$ as

$$\begin{aligned}\mathrm{Var}(X) &= \mathrm{E}(X^2) - \{\mathrm{E}(X)\}^2 \\ &= \mathrm{E}\{X(X-1)\} + \mathrm{E}(X) - \{\mathrm{E}(X)\}^2 \\ &= n(n-1)p^2 + np - n^2p^2 = np(1-p).\end{aligned}$$

Thus the expectation of X is np and its variance is $np(1-p)$.

In the numerical part of the problem we denote the numbers of defectives in the first and second samples by X_1 and X_2 respectively. These random variables are independent, and both have the binomial distribution with index $n = 6$ and parameter $p = 0{\cdot}2$, where p is the probability that an item is defective.

A batch will be accepted if $X_1 \leq 1$ or if $X_1 = 2$ and $X_2 = 0$. Thus

$$\begin{aligned}\Pr(\text{batch is accepted}) &= \Pr\{(X_1 \leq 1) \cup (X_1 = 2 \cap X_2 = 0)\} \\ &= \Pr(X_1 \leq 1) + \Pr(X_1 = 2 \cap X_2 = 0),\end{aligned}$$

by the addition law of probability for mutually exclusive events. Now

$$\begin{aligned}\Pr(X_1 \leq 1) &= \Pr(X_1 = 0) + \Pr(X_1 = 1) \\ &= (0{\cdot}8)^6 + \{6 \times (0{\cdot}2) \times (0{\cdot}8)^5\} = 0{\cdot}6554.\end{aligned}$$

We also obtain, by independence of X_1 and X_2,

$$\Pr(X_1 = 2 \cap X_2 = 0) = \left\{\binom{6}{2} \times (0{\cdot}2)^2 \times (0{\cdot}8)^4\right\} \times (0{\cdot}8)^6 = 0{\cdot}0644.$$

Therefore

$$\Pr(\text{batch is accepted}) = 0{\cdot}6553 + 0{\cdot}0644 = 0{\cdot}720.$$

Notes

(1) In the solution, we obtained $\text{Var}(X)$ by first finding an expression for $E\{X(X-1)\}$. Use of this device is far from obvious, but its value is seen from the cancellation that was possible in the binomial coefficient. In this way we could say that the device used matches the problem. The same method also works for the Poisson distribution (see Problem 2A.7), but in general the only advice that can be given is to look at the mathematical structure of the probability function and see if there is any feature which can be exploited. This is done, for example, in Note 1 to Problem 1B.2 in which the function $(X-7)^2$ is used. See also the negative binomial distribution, in Problem 2A.10, for which the variance is most easily found through the formula

$$\text{Var}(X) = E\{X(X+1)\} - E(X) - \{E(X)\}^2.$$

(2) In answering the numerical part of the problem it is necessary to assume from the statement in the problem that the batch is 'large', so that the proportion of defective articles remains constant even after sampling some of the articles without replacement. This allows us to use the binomial distribution. Strictly speaking, of course, when items are taken from a finite population, the proportion with some attribute changes; the distribution of the number of 'successes' is no longer binomial, but hypergeometric. (See Problem 2A.11 for a discussion of this distribution.)

(3) 'Success' and 'failure' are only really standard labels used for the two possible outcomes of a Bernoulli trial. In fact 'success' may sometimes constitute what we might more readily think of as failure. In the above example, for instance, we have taken 'success' to correspond to a defective article.

* (4) The technique of *probability generating functions*, though more advanced, can be used profitably in theoretical work on the binomial and other discrete distributions. In summary, a discrete distribution can be described completely by its probability generating function (or p.g.f.) just as it can by the more usual probability function. The p.g.f. of a random variable X is defined as a function $G_X(t) = E(t^X)$, and we outline some of its main properties.

(a) Expanding $G_X(t)$ as a power series in t, the coefficient of t^x is $\Pr(X = x)$; hence the name of the function.

(b) The moments (in particular the mean and variance) of X can be found by differentiating $G_X(t)$ w.r.t. t, and then evaluating the derivatives at $t = 1$. Specifically,

$$E(X) = G'_X(1)$$

and

$$E\{X(X-1)\} = G''_X(1),$$

so that

$$\text{Var}(X) = G''_X(1) + G'_X(1) - \{G'_X(1)\}^2.$$

(c) The p.g.f. of the sum of any number of independent random variables is the product of their individual p.g.fs.

Applying the technique to the total number of successes in n independent trials, in each of which the probability of success is p, we define random variables $Y_1, Y_2, \ldots, Y_n$, with Y_i being 1 if trial i is a success and 0 if not. Denoting the p.g.f. of Y_i by $G_i(t)$, we find

$$G_i(t) = (1-p)t^0 + pt^1 = (1-p)+pt, \; i = 1, 2, \ldots, n.$$

But $X = Y_1 + Y_2 + \ldots + Y_n$, so by property (c) above we obtain

$$G_X(t) = \{(1-p) + pt\}^n$$

as the p.g.f. of X, where $X \sim B(n, p)$.

We can now use this result to obtain the mean and variance of X. Differentiating, we obtain

$$G'_X(t) = np\{(1-p) + pt\}^{n-1},$$

and substituting $t = 1$ gives, immediately, $E(X) = np$. Differentiating again, we find that the second derivative is, similarly, $n(n-1)p^2$ when $t = 1$, and thus we obtain, easily,

$$\text{Var}(X) = n(n-1)p^2 + np - (np)^2 = np(1-p).$$

Note that by expanding the p.g.f. in powers of t we obtain, as the term in t^x,

$$\binom{n}{x}(1-p)^{n-x}p^x t^x,$$

and thus find the coefficient of t^x to be the probability function of the binomial distribution.

2A.3 Sampling for defective items

A machine produces articles of which an average of 10% are defective. Find an approximate value for the probability that a random sample of 500 of these articles contains more than 25 which are defective. What, approximately, is the probability that the sample contains fewer than 60 defectives?

Solution

The number of defectives, X, in the random sample of size 500 has the binomial distribution with $n = 500$ and $p = 0{\cdot}1$. To obtain an approximate value for the probability $\Pr(X > 25)$, we use the normal approximation to this distribution: that is, we use the normal distribution with the same mean and variance. We have

$$\text{mean} = np = 500 \times 0{\cdot}1 = 50, \text{ variance} = np(1-p) = 500 \times 0{\cdot}1 \times 0{\cdot}9 = 45.$$

Hence we use the $N(50, 45)$ distribution or, equivalently, transform to

$$Z = \frac{X - 50}{\sqrt{45}}$$

and use the standardised normal distribution $N(0, 1)$, with distribution function $\Phi(z)$.

Since we are approximating a discrete distribution (binomial) by a continuous distribution (normal) we include a continuity correction. From tables of the normal distribution we find that

$$\Pr(X \leq 25) \approx \Phi\left(\frac{25+\frac{1}{2}-50}{6{\cdot}7082}\right) = \Phi(-3{\cdot}652) = 1-0{\cdot}9999 = 0{\cdot}0001.$$

But we require $\Pr(X > 25)$, given by $\Pr(X > 25) = 1 - \Pr(X \leq 25) = 0{\cdot}9999$.

Similarly, for $\Pr(X < 60)$ we require

$$\Phi\left(\frac{60-\frac{1}{2}-50}{6{\cdot}7082}\right) = \Phi(1{\cdot}416) = 0{\cdot}9216.$$

The probability that there are fewer than 60 defectives is thus approximately 0·922.

Notes

(1) As stated in the solution, the distribution of the number of defectives in the sample is binomial. This is because, for each of the 500 articles separately, whether or not it is found to be defective can be thought to constitute a Bernoulli trial, independent of all the others. (See the Note to Problem 2A.1.) However, since the number of articles in the sample is large, the normal approximation can be safely used here; the words 'approximate' and 'approximately' in the statement of the problem indicate that this is what is required. Note that the normal distribution can provide a good approximation to the binomial distribution even when this is not symmetrical, as will be the case when $p \neq \frac{1}{2}$. However, the further from $\frac{1}{2}$ the value of p is, the larger will be the value of n required for the approximation to be a good one.

A rough check on whether or not the normal distribution is likely to provide a good approximation to a particular binomial distribution can be carried out by seeing how much of the tails of the normal distribution is outside the range of the binomial distribution (a binomially distributed random variable has a finite range, whereas that of a normally distributed random variable is infinite). For example, in the above case, 0 is closer to the mean (50) than 500, but even this is more than 7 standard deviations away from the mean. Since much less than 1% of the normal distribution is contained beyond 3 standard deviations from the mean, the indication here is that the normal approximation should be very good. A useful rule of thumb is that the approximation is reasonable if both np and $n(1-p)$ exceed 5.

(2) A continuity correction is used because a discrete distribution is being approximated by a continuous distribution. To see what is happening here, one can think of approximating the probability that $X = 25$, for instance, in this problem. Now, for a random variable with a continuous distribution, the probability that the variable takes some specified value *exactly* is zero. This is clearly not the case for a random variable which has a discrete distribution. The above probability is then approximated by the area under the probability density function curve for $N(50, 45)$ between the values 24·5 and 25·5. Similarly the probability that $X = 24$ is approximated by the area under the curve between 23·5 and 24·5; continuing in this way, the probability that $X \leq 25$ is approximated by the entire area to the left of 25·5.

(3) The final probability in the solution has been rounded to 3 decimal places because linear interpolation in the tables of the normal distribution was used to obtain the fourth figure, and since the calculated probability is only an approximation anyway.

2A.4 The music recital

Regular music recitals are held in a small hall with seating for an audience of 98 people. The booking office staff find that, on average, 3% of people who book for a recital fail to turn up, and adopt a policy of selling up to 100 tickets for any recital. What is the probability that for a recital for which 100 tickets have been sold, everyone who turns up has a seat?

Solution

The number X of ticketholders who fail to turn up for the recital has a binomial distribution with index $n = 100$ and parameter $p = 0{\cdot}03$, and hence mean $= np = 3$. We need to find the probability that two or more people fail to turn up (when there will be a seat for everyone), i.e. to find

$$\Pr(X \geq 2) = 1 - \Pr(X < 2) = 1 - \Pr(X = 0) - \Pr(X = 1).$$

Since n is large and p is small we may approximate the binomial probabilities using the Poisson distribution with the same mean, 3. Hence the required probability is (approximately)

$$1 - e^{-3} - 3e^{-3} = 1 - 4e^{-3} = 1 - 0{\cdot}199 = 0{\cdot}801.$$

Notes

(1) We are told that, on average, 3% of people who have booked for recitals fail to turn up, and have to make the assumption that the rate of non-attendance stays constant over time when computing the required probability. In practice this may not be the case as, for example, different times of the year or the varying fame of performers may affect the attendance rates. We are also, in using the binomial distribution, making the assumption that people who have booked act independently with regard to attendance or non-attendance. It is most unlikely that this assumption is valid, since many will book in small groups and it will often be the case that the failure of one member of a group to attend will be associated with the non-attendance of some or all of the other members of that group. Thus a solution based on the binomial distribution can only be approximate. It is clear, however, that such a solution is the best possible given the information at our disposal.

(2) The use of the Poisson approximation to the binomial distribution is less necessary now than it was when the only computational aids generally available were mathematical tables and mechanical calculators. With a modern scientific calculator it is (almost) a simple matter to calculate the required binomial probabilities exactly, as follows:

$$\Pr(X \geq 2) = 1 - (0{\cdot}97)^{100} - 100 \times (0{\cdot}03) \times (0{\cdot}97)^{99} = 0{\cdot}805.$$

The reason we say that the direct calculation of binomial probabilities is *almost* a simple matter is that, if n is large, calculation of the necessary binomial coefficients by substituting the values of n and x in

$$\binom{n}{x} = \frac{n!}{x!(n-x)!}$$

will fail because some of the factorials involved will overflow the calculator. This may be avoided, as in the above calculations, by appropriate cancellation before entering any figures into the calculator. It may also be avoided by calculating probabilities sequentially, as is done for the Poisson distribution in Problem 2A.8: beginning with the value of $\Pr(X = 0)$, we obtain the probabilities of other values of X using the recurrence relationship

$$\Pr(X = x) = \frac{n - x + 1}{x}\frac{p}{1-p}\Pr(X = x - 1).$$

(3) Confusion often arises as to which approximation to the binomial distribution to use when n is large. The two possibilities are the Poisson and normal distributions. The Poisson distribution is, mathematically, the limiting form of $B(n,p)$ as $n \to \infty$ and $p \to 0$ in such a way that the product np remains constant. The present problem gives an example where n is large and p is small, and hence the Poisson approximation is appropriate.

The normal distribution approximation also depends mathematically on n tending to infinity, but note the rule of thumb given in Note 1 to Problem 2A.3 that the approximation is likely to be reasonable if the products np and $np(1-p)$ are both greater than 5. (It is also worth noting that the normal distribution can be derived as a limiting form of the Poisson as its single parameter, μ, tends to infinity, so that the normal distribution can be used to approximate the Poisson distribution for large μ. We thus see how the Poisson approximation to the binomial distribution might in turn lead to a normal approximation.)

Although we see from this discussion that it is clearly the Poisson distribution which is appropriate to the present problem, it makes little difference numerically which approximation is used. If we use the normal approximation we find the answer to be 0·810 rather than 0·801. However, this does not alter the fact that, from the structure of the problem, it is the Poisson, not the normal, distribution which provides the appropriate approximation.

2A.5 Marking a multiple-choice test

(a) A multiple-choice test paper contains 50 questions; for each question three answers are given, one of which is correct. The two incorrect answers to any question are designed to be plausible, so that an ignorant candidate could be expected to pick an answer quite at random. If the examination is marked simply by giving one mark per correct answer, what should the pass mark be if the probability that a completely ignorant candidate passes is to be approximately 1%?

(b) Suppose now that the examination is marked by awarding two marks per correct answer, but deducting one mark for every incorrect answer. If an ignorant candidate attempts every question, what is the expectation and variance of the candidate's total score?

(c) Consider the position of a candidate when the scoring system is as in part (b), but with only one mark for a correct answer, and when the pass mark is 28. The candidate has revised half the syllabus thoroughly, and finds that he is certain of the correct answers to half the questions. To gain the extra 3 marks needed to pass he decides to guess at the answers to a few more questions. Would the probability of passing be greater if he picked just three more questions hoping to get them all correct, or if he guessed at five questions?

Solution

(a) Let the number of correct answers given by an ignorant candidate be denoted by X. Then $X \sim B(50, \frac{1}{3})$, and for this distribution a normal approximation is acceptable, so we conclude that, approximately,

$$X \sim N(50\times\tfrac{1}{3}, 50\times\tfrac{1}{3}\times\tfrac{2}{3}), \text{ i.e. } X \sim N(\tfrac{50}{3}, (\tfrac{10}{3})^2).$$

We are required to find that value x such that $\Pr(X \geq x) \approx 0{\cdot}01$ and so we have to solve for x in the equation

$$\Pr(X \geq x) \approx 1 - \Phi\left(\frac{x - \frac{1}{2} - \frac{50}{3}}{\frac{10}{3}}\right) = 0{\cdot}01.$$

Now $\Phi(k) = 0{\cdot}99$ implies that $k = 2{\cdot}3263$, so we have

$$\frac{x - \frac{1}{2} - \frac{50}{3}}{\frac{10}{3}} = 2{\cdot}3263,$$

or $x = \dfrac{1}{2} + \dfrac{50}{3} + \dfrac{23{\cdot}263}{3} = 24{\cdot}92$. In practice this obviously means that the pass mark should be set at 25.

(b) As in part (a), the distribution of X is $B(50, \frac{1}{3})$. Since X is the number of correct answers and $50-X$ the number of incorrect answers, the score S of the candidate who attempts every question is a random variable such that

$$S = 2X + (-1)(50-X) = 3X - 50.$$

We see therefore that $E(S) = 3E(X) - 50$ and $Var(S) = 9Var(X)$. From standard properties of the binomial distribution used in part (a), $E(X) = \frac{50}{3}$ and $Var(X) = \frac{100}{9}$, so we reach

$$E(S) = 0, \quad Var(S) = 100.$$

(c) With one mark per correct answer and half the syllabus known, the candidate has 25 marks in the bag, with up to 25 further questions available to secure the extra 3 marks. (The problem with guessing at many of these 25 further questions is that he is twice as likely to guess incorrectly, and be penalised, as to get an answer right.) If three further questions are attempted, all must be correct in order to pass, and the probability p_3 of this is $p_3 = (\frac{1}{3})^3$. If five questions are attempted, then at least 4 must be correct to secure the three marks needed, and the probability p_5 of this is

$$p_5 = (\tfrac{1}{3})^5 + 5\times(\tfrac{1}{3})^4(\tfrac{2}{3}) = \frac{11}{3^5}.$$

Since $p_3 = \frac{9}{3^5} < \frac{11}{3^5} = p_5$, we see that the better strategy is to try five further questions.

Notes

(1) While the numbers have been kept simple, the set-up is a moderately realistic one, and scoring systems of the type described are in use. They do, of course, offer candidates considerable incentive to revise, since guesswork is most unlikely to be rewarded. In practice, candidates are usually not guessing quite at random, and if the scoring scheme is known to them they have interesting strategic choices. For example, to maximise the total expected score, a candidate should attempt a question if he assesses the chance of answering it correctly at $\frac{1}{3}$ or more under the scheme in (b), but under the system in (c) only if the chance exceeds $\frac{1}{2}$.

Other possibilities include allowing candidates to check two (or, if appropriate, even more) answers to the same question, and gain reduced credit if the correct answer is one of those checked.

(2) In part (c) the statement of the problem allows us to restrict attention to just 3 or 5 extra questions. In practice one would wish to calculate the optimum number of extra questions to choose. A little reasoning shows that choosing four questions is not sensible, since the candidate can then pass only by answering all four correctly, a more difficult task than guessing three out of three. Similarly, choosing six questions is worse than choosing five. However, a little calculation shows that, if seven extra questions are attempted, the probability p_7 of getting five or more correct is $11/3^5$, and this strategy is thus just as good as choosing five extra questions. Trying more than seven questions is not advisable, though. If eight or nine questions are tried, then at least six must be answered correctly. Trying nine questions is therefore a better strategy than trying eight, and the probability of success from nine questions is $835/3^9$, which is smaller than p_5 and p_7.

2A.6 The insect breeding experiment

(a) Two sets of n independent trials are performed, independently of each other, and each trial results in either success or failure, the probability of success being p_1 in the first set of trials and p_2 in the second set. Show that the probability P of obtaining x_1 successes in the first set and x_2 successes in the second set is given by

$$P = Kp_1^{x_1}p_2^{x_2}(1-p_1)^{n-x_1}(1-p_2)^{n-x_2},$$

where K depends only on n, x_1 and x_2. If $p_1 = p$ and $p_2 = p^2$, find an expression for $\log P$ and show that, for given values of n, x_1 and x_2, $\log P$ has a maximum value when p is such that

$$(x_1 + 2x_2) - (n - x_1)p - 3np^2 = 0.$$

(b) An insect breeding experiment was conducted in two sections, in each of which 100 insects of a particular species were raised. In the first section a proportion p was expected to have a certain colour variation and in the second section the proportion with this colour variation was expected to be p^2, but the value of p was not known. In the event there were 22 insects in the first section, and 7 in the second section, which possessed the colour variation. Find the value of p which maximises the probability of this result.

Solution

(a) The number of successes in the first set of n independent trials has a binomial distribution, and the probability of obtaining x_1 successes is

$$\binom{n}{x_1} p_1^{x_1}(1-p_1)^{n-x_1}.$$

Similarly, the probability of obtaining x_2 successes in the second set of n independent trials is given by the binomial distribution as

$$\binom{n}{x_2} p_2^{x_2}(1-p_2)^{n-x_2}.$$

The two sets of n independent trials are also independent of each other, so that the two probabilities above should be multiplied together to yield the probability of the composite event described in the statement of the problem. The required probability is therefore

$$P = Kp_1^{x_1}p_2^{x_2}(1-p_1)^{n-x_1}(1-p_2)^{n-x_2},$$

where

$$K = \binom{n}{x_1}\binom{n}{x_2};$$

clearly K depends only on n, x_1 and x_2.

Putting $p_1 = p$ and $p_2 = p^2$ in the expression for P, we obtain

$$P = Kp^{x_1+2x_2}(1-p)^{n-x_1}(1-p^2)^{n-x_2}$$

and, taking logarithms,

$$\log P = \log K + (x_1+2x_2)\log p + (n-x_1)\log(1-p) + (n-x_2)\log(1-p^2).$$

To maximise $\log P$, whilst holding n, x_1 and x_2 constant, we differentiate with respect to p to obtain

$$\frac{\mathrm{dlog}\, P}{\mathrm{d}p} = \frac{x_1+2x_2}{p} - \frac{n-x_1}{1-p} - \frac{2p(n-x_2)}{1-p^2}.$$

Equating this to zero leads to the equation

$$(x_1+2x_2)(1-p^2) - (n-x_1)p(1+p) - 2(n-x_2)p^2 = 0,$$

$$\text{i.e. } (x_1+2x_2) - (n-x_1)p - 3np^2 = 0,$$

as required. A local maximum value of $\log P$ is then given when the value of p is one of the roots of this quadratic equation. Now denoting the left hand side of the equation in p above by $g(p)$, we see that $g(0) > 0$ and that $g(1) < 0$; also that the coefficient of p^2 is negative. It follows that the roots of the equation $g(p) = 0$ are real, and that one lies in the range $(0, 1)$ and the other is negative. (There are special cases when $x_1 = x_2 = 0$ and when they both equal n.) Differentiation shows that the second derivative of $\log P$ is negative whenever $p > 0$, so the root of the equation $g(p) = 0$ in $(0, 1)$ gives a maximum for $\log P$.

(b) We note that the situation in the insect breeding experiment is precisely that which was defined in part (a), but with particular values of n (100), x_1 (22) and x_2 (7) specified. We therefore substitute for n, x_1 and x_2 in the equation for p, and obtain

$$36 - 78p - 300p^2 = 0, \text{ or } 50p^2 + 13p - 6 = 0.$$

This factorises to give

$$(25p - 6)(2p + 1) = 0,$$

which has solutions $p = \frac{6}{25}$ and $p = -\frac{1}{2}$. Since p is a probability, between 0 and 1, the root that we require is $p = \frac{6}{25}$. The general result given in part (a) shows that this root maximises the value of $\log P$.

Notes

(1) The second, numerical, part of the problem follows directly from the first, algebraic, part. When answering examination questions, candidates often fail to recognise the relevance of the first part of a question to the second part.

(2) Although it is not necessary to realise it in order to solve the problem, the method being used here to find the value of p maximising $\log P$ is the method of *maximum likelihood estimation*. The model for the insect breeding experiment depends on a parameter p, which we wish to estimate. The method proceeds by finding that value of p for which the probability that the values actually observed occur is maximised. The method of maximum likelihood is one of the most commonly used methods of estimation in statistics (see Problems 2A.11 and 4D.4 for other examples of its use). Another is the method of least squares (see Note 4 to Problem 5A.1 for its use in the context of linear regression). Others are the method of moments and the minimum chi-squared method.

(3) In applying the method of maximum likelihood to a problem like the present one, we usually find that the quantity to be maximised is a product of several terms (themselves probabilities or probability densities). It is usually easier to find the maximum of a *sum* of terms rather than that of a *product*. A simple way to convert a sum to a product is to take logarithms, as was done here. Now since $\log P$ is a monotonic function of P, if we require the value of p maximising P, this can equivalently, and more easily, be done by finding the value maximising $\log P$. This is what was done in the solution above.

2A.7 The telephone exchange

A telephone exchange receives, on average, 5 calls per minute. Find the probability

(i) that in a 1-minute period no calls are received;

(ii) that in a 2-minute period fewer than 4 calls are received;

(iii) that in a 20-minute period no more than 102 calls are received;

(iv) that out of five separate 1-minute periods there are exactly four in which 2 or more calls are received.

Solution

(i) We assume that the number of incoming calls in one minute has the Poisson distribution with mean 5. Hence, if X is the number of incoming calls in one minute,

$$\Pr(X = x) = e^{-5}\frac{5^x}{x!}, \; x = 0, 1, 2, \ldots.$$

Then

$$\Pr(X = 0) = e^{-5} = 0{\cdot}007.$$

(ii) The number of calls in a 2-minute period, Y, has the Poisson distribution with mean 10. Hence

$$\Pr(Y < 4) = \Pr(Y = 0, 1, 2 \text{ or } 3)$$

$$= e^{-10}\left(1 + 10 + \frac{10^2}{2!} + \frac{10^3}{3!}\right) = 0{\cdot}010.$$

(iii) The number of calls, W, in 20 minutes has the Poisson distribution with mean 100. We require $\Pr(W \leq 102)$ and, since the mean of the distribution is large, are able to use a normal approximation. Since the mean and variance of the Poisson distribution are the same, the appropriate normal distribution is $N(100, 10^2)$. Using a continuity correction, we find

$$\Pr(W \leq 102) \approx \Phi\left(\frac{102 + \frac{1}{2} - 100}{10}\right)$$

$$= \Phi(0{\cdot}25) = 0{\cdot}599.$$

(iv) In any 1-minute period chosen at random the probability of 2 or more calls being received is

$$1 - \Pr(0 \text{ or } 1 \text{ calls}) = 1 - e^{-5}(1 + 5)$$

$$= 0{\cdot}9596.$$

Then, using the binomial distribution $B(5, 0{\cdot}9596)$, the probability that exactly four out of five 1-minute periods contain 2 or more calls is

$$5(0{\cdot}9596)^4(1 - 0{\cdot}9596) = 0{\cdot}171.$$

Notes

(1) Although it is not explicitly stated in the problem, the intention is that we should assume that the number of telephone calls in one minute has a Poisson distribution. This arises if we assume that calls are distributed at random over time, with no tendency either to arrive in groups or to be evenly spaced. We also assume that there is no change over time in the rate at which calls are received.

A mathematically precise formulation of these assumptions is provided by the *Poisson process*, a model for completely random occurrences in time, with the following properties:

(i) events relating to non-overlapping intervals of time are statistically independent,

(ii) in any small time interval $(t, t + \delta t)$, the probability of an occurrence in the interval is proportional to the length of the interval,

(iii) in any small time interval $(t, t + \delta t)$, the probability of two or more occurrences is proportional to $(\delta t)^2$, i.e. is negligible.

While we have given the exact definition, for completeness, the important fact is simply that, when occurrences are completely haphazard in time, the number of these occurrences in a fixed time is a random variable with a Poisson distribution. It is this which justifies use of the Poisson distribution in so many cases involving accidents, infrequent occurrences, etc. An application of the Poisson process to the arrival of customers at a queue can be found in Problem 4D.4. Similar arguments, with 'space' replacing 'time', justify the appropriateness of the Poisson distribution in other contexts (see, for example, Problem 2A.8).

In many applications, but in particular in the field of ecology, the Poisson distribution is often used to describe the distribution of organisms in two- or three-dimensional space. Alternative distributions (if the organisms are not distributed randomly) can be described as regular or clustered, and interest often centres on fitting one of the so-called *contagious* distributions to clustered, clumped or aggregated data. An example of a contagious distribution is the negative binomial distribution (see Problem 2A.10); all have the property that the variance is greater than the mean.

(2) In part (iv) we need to compute probabilities from two distributions (Poisson and binomial), and to combine these. It is not uncommon to find problems of this type, where the probability p of 'success' in a Bernoulli trial is calculated from some standard distribution, and then used in the calculation of a binomial probability.

2A.8 Random sultanas in scones

(a) Show that for a Poisson distribution the mean is equal to the variance.

(b) In a bakery 3600 sultanas are added to a mixture which is subsequently divided up to make 1200 fruit scones. Assuming that the number of sultanas in each scone follows a Poisson distribution, estimate

(i) the number of scones which will be without a sultana;

(ii) the number with 5 or more sultanas.

Solution

(a) If $X \sim \text{Poisson}(\mu)$, then $\Pr(X = x) = e^{-\mu}\mu^x/x!$, $x = 0, 1, \ldots$, and so

$$\begin{aligned} E(X) &= \sum_{x=0}^{\infty} x \Pr(X = x) \\ &= \sum_{x=1}^{\infty} x \Pr(X = x) = e^{-\mu} \sum_{x=1}^{\infty} x \mu^x / x! \\ &= \mu e^{-\mu} \sum_{x=1}^{\infty} \mu^{x-1}/(x-1)! = \mu, \end{aligned}$$

since substituting $y = x - 1$ in the summation shows it to be just the expansion of e^{μ} in powers of μ.

For Var(X), we first consider $E\{X(X-1)\}$. By an essentially similar argument, this is seen to be

$$\begin{aligned} E\{X(X-1)\} &= \sum_{x=2}^{\infty} x(x-1) e^{-\mu}\mu^x/x! \\ &= \mu^2 e^{-\mu} \sum_{x=2}^{\infty} \mu^{x-2}/(x-2)! = \mu^2, \end{aligned}$$

this time using a substitution $y = x - 2$ in the summation. We now find

$$\begin{aligned} \text{Var}(X) &= E\{X(X-1)\} + E(X) - \{E(X)\}^2 \\ &= \mu^2 + \mu - \mu^2 = \mu. \end{aligned}$$

We see therefore that for the Poisson distribution the mean and variance are equal.

(b) The mean of the Poisson distribution is simply the average number of sultanas per scone, which is 3600/1200 = 3. Hence, if X is the number of sultanas in a scone,

$$\Pr(X = x) = e^{-3}\frac{3^x}{x!}, \; x = 0, 1, \ldots.$$

Substituting $x = 0$ in this expression, we obtain the probability that any one scone contains no sultanas as $e^{-3} = 0{\cdot}049\,79$. To estimate the number of scones without a sultana we need to multiply this probability by the total number of scones. Hence, since $1200 \times 0{\cdot}049\,79 = 59{\cdot}7$, the estimated number of scones without a sultana is 60.

Now

$$\Pr(5 \text{ or more sultanas}) = 1 - \sum_{x=0}^{4} \Pr(X = x).$$

Computing the required probabilities sequentially, using the recurrence relationship

$$\Pr(X = x) = \frac{3}{x}\Pr(X = x - 1),$$

we obtain

$$\Pr(X = 1) = \tfrac{3}{1}\Pr(X = 0) = 0{\cdot}1494;$$

$$\Pr(X = 2) = \tfrac{3}{2}\Pr(X = 1) = 0{\cdot}2240;$$

$$\Pr(X = 3) = \tfrac{3}{3}\Pr(X = 2) = 0{\cdot}2240;$$

$$\Pr(X = 4) = \tfrac{3}{4}\Pr(X = 3) = 0{\cdot}1680.$$

Hence $\sum_{x=0}^{4} \Pr(X = x) = 0{\cdot}8153$ and so

$$\Pr(5 \text{ or more sultanas}) = 1 - 0{\cdot}8153 = 0{\cdot}1847.$$

The estimated number of scones with 5 or more sultanas is therefore $1200 \times 0{\cdot}1847 = 221{\cdot}7$, i.e. 222 to the nearest whole number.

Notes

(1) In this problem the use of the Poisson distribution depends on an assumption that the sultanas are 'randomly' distributed through the mixture. For a detailed discussion of what the word 'randomly' means in this context, see Note 1 to Problem 2A.7.

(2) To find values of the Poisson probability function in solving this problem, we have made use of a simple relationship between successive values. Such an approach can frequently simplify calculations when we are dealing with discrete distributions (see, for example, Note 2 to Problem 2A.4). A potential hazard here, though, is that of compounding rounding errors. Care needs to be taken to account for a sufficiently large number of decimal places in the working to achieve the degree of accuracy ultimately desired. Better still, if a suitable calculator is available, is to carry out the sequence of calculations entirely within the machine, writing down any results needed but basing subsequent calculations on the more accurate value still held in the calculator.

The reader may have noted that this was done in the calculation above. The probability that $X = 1$ was recorded as $0{\cdot}1494$ while $\Pr(X = 2)$ was recorded as $0{\cdot}2240$. But $\tfrac{3}{2} \times 0{\cdot}1494 = 0{\cdot}2241$, yet in fact both figures given in the solution are accurate to four decimal places.

(3) Note that $\Pr(X = 2) = \Pr(X = 3)$ in this example. This illustrates a property of the Poisson distribution which holds whenever the mean μ is an integer, namely that $\Pr(X = \mu - 1) = \Pr(X = \mu)$. When μ is not an integer, then if $\mu > 1$ the successive probabilities increase progressively to a unique maximum (corresponding to the largest integer less than μ), and then decrease monotonically, while if $\mu < 1$ they simply decrease monotonically.

(4) In calculating $\mathrm{Var}(X)$ in part (a) we used the device of first calculating $E\{X(X - 1)\}$. A short discussion of this device can be found in Note 1 to Problem 2A.2.

* (5) The problem of obtaining the mean and variance of a Poisson distribution can be tackled using the probability generating function. (The function is defined generally, and some of its properties are given, in Note 4 to Problem 2A.2.)

If a random variable X has a Poisson distribution with mean μ, it has p.g.f. $G_X(t)$ given by

$$
\begin{aligned}
G_X(t) = \mathrm{E}(t^X) &= \sum_{x=0}^{\infty} t^x \Pr(X = x) \\
&= \sum_{x=0}^{\infty} t^x \frac{\mathrm{e}^{-\mu}\mu^x}{x!} \\
&= \mathrm{e}^{-\mu} \sum_{x=0}^{\infty} \frac{(t\mu)^x}{x!} \\
&= \mathrm{e}^{-\mu}\mathrm{e}^{t\mu} = \mathrm{e}^{\mu(t-1)}.
\end{aligned}
$$

In the present problem we require the mean and variance of X, and these can be found by differentiation. For the mean we require $G'_X(1)$, and obtain

$$G'_X(t) = \mu \mathrm{e}^{\mu(t-1)},$$

and evaluating this at $t = 1$ gives $\mathrm{E}(X) = \mu$. Differentiating again, we find the second derivative to be μ^2 when $t = 1$, so, from the result presented in Note 4 to Problem 2A.2, we obtain

$$\mathrm{Var}(X) = G''_X(1) + G'_X(1) - \{G'_X(1)\}^2 = \mu^2 + \mu - \mu^2 = \mu = \mathrm{E}(X).$$

* 2A.9 Relationship between binomial and Poisson distributions

Independent random variables X and Y have Poisson distributions with means μ_1 and μ_2 respectively, and the random variable Z is defined as $Z = X + Y$.

(a) Show that the distribution of Z is a Poisson distribution.

(b) Show that, conditional upon the event $Z = z$, the distribution of X is binomial, and find its index and parameter.

Solution

(a) The most direct method evaluates $\Pr(Z = z)$, for values of $z \geq 0$, by splitting the event $\{Z = z\}$ into mutually exclusive component events whose probabilities can then be added together. Now $Z = z$ if $X = 0$ and $Y = z$, or if $X = 1$ and $Y = z-1$, or if $X = 2$ and $Y = z-2$, and so on, ending with the event that $X = z$ and $Y = 0$. These $z + 1$ components are mutually exclusive, and each is the intersection of two independent events, for example $\{X = 1\}$ and $\{Y = z-1\}$. We thus obtain

$$\Pr(Z = z) = \sum_{x=0}^{z} \Pr(X = x \cap Y = z-x),$$

by the addition law of probability. Further, by independence,

$$
\begin{aligned}
\Pr(X = x \cap Y = z-x) &= \Pr(X = x)\Pr(Y = z-x) \\
&= \frac{\mathrm{e}^{-\mu_1}\mu_1^x}{x!} \frac{\mathrm{e}^{-\mu_2}\mu_2^{z-x}}{(z-x)!},
\end{aligned}
$$

and so

$$\Pr(Z = z) = \sum_{x=0}^{z} \frac{\mathrm{e}^{-\mu_1}\mu_1^x}{x!} \frac{\mathrm{e}^{-\mu_2}\mu_2^{z-x}}{(z-x)!}.$$

Rearranging the terms in this expression gives

$$\Pr(Z=z) = e^{-(\mu_1+\mu_2)} \sum_{x=0}^{z} \frac{1}{x!\,(z-x)!} \mu_1^x \mu_2^{z-x}$$

$$= \frac{e^{-(\mu_1+\mu_2)}(\mu_1+\mu_2)^z}{z!} \sum_{x=0}^{z} \frac{z!}{x!\,(z-x)!} \left(\frac{\mu_1}{\mu_1+\mu_2}\right)^x \left(\frac{\mu_2}{\mu_1+\mu_2}\right)^{z-x}$$

$$= \frac{e^{-(\mu_1+\mu_2)}(\mu_1+\mu_2)^z}{z!} \sum_{x=0}^{z} \binom{z}{x} p^x (1-p)^{z-x},$$

where, for clarity, we have written $p = \mu_1/(\mu_1+\mu_2)$. It is clear that the sum in this expression is just the binomial expansion of $\{p + (1-p)\}^z$, and is thus 1; hence

$$\Pr(Z=z) = \frac{e^{-(\mu_1+\mu_2)}(\mu_1+\mu_2)^z}{z!}, \quad z = 0, 1, 2, \ldots,$$

i.e. Z has the Poisson distribution with mean $\mu_1+\mu_2$.

(b) To find the conditional distribution of X given the event $Z=z$, we simply need to evaluate $\Pr(X=x \mid Z=z)$, using the ordinary formula for conditional probability

$$\Pr(X=x \mid Z=z) = \frac{\Pr(X=x \cap Z=z)}{\Pr(Z=z)}.$$

The denominator was found in part (a), and the numerator can be re-expressed as $\Pr(X=x \cap Y=z-x)$. The advantage of so doing is that X and Z are not independent, but X and Y are, so that the joint probability can be evaluated as a product, viz.

$$\Pr(X=x \cap Z=z) = \Pr(X=x \cap Y=z-x) = \frac{e^{-\mu_1}\mu_1^x}{x!} \frac{e^{-\mu_2}\mu_2^{z-x}}{(z-x)!};$$

we thus obtain

$$\Pr(X=x \mid Z=z) = \frac{\dfrac{e^{-\mu_1}\mu_1^x}{x!} \dfrac{e^{-\mu_2}\mu_2^{z-x}}{(z-x)!}}{e^{-(\mu_1+\mu_2)}(\mu_1+\mu_2)^z/z!}$$

$$= \frac{z!}{x!\,(z-x)!} \left(\frac{\mu_1}{\mu_1+\mu_2}\right)^x \left(\frac{\mu_2}{\mu_1+\mu_2}\right)^{z-x};$$

i.e. conditional upon $Z=z$, $X \sim B\left(z, \dfrac{\mu_1}{\mu_1+\mu_2}\right)$.

Notes

(1) This rather tricky problem is included partly to provide a basis for a later one, Problem 4C.8. (While the present problem is purely an exercise in manipulating probabilities, its application is quite practical, providing us with a two-sample test for use with data from Poisson distributions.) The problem also shows how closely interlinked the binomial and Poisson distributions are; it is not just the case that one can sometimes be used as an approximation to the other.

(2) As in Problem 2A.2, probability generating functions (p.g.fs) can be used to advantage here, since the aim in part (a) is to find the distribution of the sum of two independent random variables. (Further details of definition, etc. can be found in Note 4 to Problem 2A.2.) If a random variable W has a Poisson distribution with mean μ, it has p.g.f. $G_W(t)$ given by

$$G_W(t) = \mathrm{E}(t^W) = \sum_{w=0}^{\infty} t^w \Pr(W = w) = e^{\mu(t-1)},$$

as seen in Problem 2A.8. Conversely, any function of t of this form will be the p.g.f. of a random variable with a Poisson distribution, and whatever replaces μ in the formula will be the mean of its distribution.

Now X and Y both have Poisson distributions, so we find

$$G_X(t) = e^{\mu_1(t-1)}$$

and $$G_Y(t) = e^{\mu_2(t-1)}.$$

But $Z = X + Y$, so, by the multiplicative property of p.g.fs,

$$G_Z(t) = G_X(t) \times G_Y(t)$$
$$= e^{(\mu_1 + \mu_2)(t-1)},$$

and from the form of this function of t we see that Z must have the Poisson distribution with mean $\mu_1 + \mu_2$.

This argument is only slightly shorter than the more direct one used in the solution. But it is much more powerful, since it extends immediately to finding the distribution of the sum of three or more independent random variables; it follows naturally that the sum of any number of independent Poisson random variables will itself have a Poisson distribution. All one has to do is to multiply together the appropriate number of p.g.fs, and the product will be seen to be of the required form. By contrast, using an extension of the direct method of the solution to part (a) is very cumbersome.

2A.10 Tossing a coin until 'heads' appears

(a) In a series of independent tosses of a coin, with probability p of the coin landing 'heads', and probability $1-p$ of it landing 'tails', obtain an expression for $\Pr(X = x)$, $x = 1, 2, 3, \ldots$, where X is a random variable denoting the number of tosses until the first head appears.

(b) If m is a positive integer, obtain an expression for $\Pr(Y = y)$, $y = m, m+1, m+2, \ldots$, where Y is a random variable denoting the number of tosses until the mth head appears.

Solution

(a) As an illustration, consider the case $x = 5$, when we observe:

TTTTH

Since the tosses are independent, we obtain

$$\Pr(X = 5) = (1-p)^4 p,$$

and, in general,

$$\Pr(X = x) = (1-p)^{x-1} p, \quad x = 1, 2, 3, \ldots.$$

We say that the random variable X has a *geometric* distribution, with parameter p.

(b) In this case, to obtain the result $Y = y$, we must have a sequence of tosses of the following form:

Toss number	1 2 3 4 5 ... $y-1$	y
Result	TTHHT ... T	H

The first $y-1$ tosses result in $m-1$ heads and $y-m$ tails, with these heads and tails arranged in any order. Thus the sequence TTHHT...T illustrated corresponds to just one of the

$\binom{y-1}{m-1}$ possible orderings for the first $y-1$ tosses in which there are exactly $m-1$ heads; the probability of obtaining $m-1$ heads from these tosses is found from the binomial $B(y-1,p)$ distribution. The probability that the yth toss is a head is simply p. We see therefore that

$$\Pr(Y=y) = \binom{y-1}{m-1}p^{m-1}(1-p)^{y-m}\times p$$
$$= \binom{y-1}{m-1}p^{m}(1-p)^{y-m}, \quad y=m, m+1, \ldots.$$

The random variable Y is said to have a *negative binomial* distribution, with parameters m and p. We note that when $m=1$ we have the special case of the geometric distribution found in part (a).

Notes

(1) We may readily check that, for the geometric distribution, the probability function sums to 1, since

$$\sum_{x=1}^{\infty}\Pr(X=x) = p\sum_{x=1}^{\infty}(1-p)^{x-1} = \frac{p}{1-(1-p)} = 1.$$

The reader may care to perform the same check for the negative binomial distribution.

* (2) By considering the number of heads in $n+m$ independent tosses of a coin we can readily verify the following result, which connects negative binomial and binomial random variables:

$$\Pr(Y \le n+m) = \Pr(Z \ge m), \qquad (*)$$

for $n = 0,1,2,\ldots,$ where Y has the negative binomial distribution given in the problem and Z is a binomial random variable with the $B(n+m,p)$ distribution.

The event $\{Z \ge m\}$ occurs if there are at least m heads in the first $n+m$ tosses, which implies that the number of tosses Y until the mth head must be no larger than $n+m$. The two events $\{Y \le n+m\}$ and $\{Z \ge m\}$ are identical, and thus have the same probability, so that equation $(*)$ holds.

It is instructive (and far more difficult) to prove equation $(*)$ using algebra. We require

$$\sum_{y=m}^{n+m}\binom{y-1}{m-1}p^{m}(1-p)^{y-m} = \sum_{z=m}^{n+m}\binom{n+m}{z}p^{z}(1-p)^{n+m-z},$$

i.e.

$$p^m\sum_{k=0}^{n}\binom{m+k-1}{k}(1-p)^k = p^m\sum_{k=0}^{n}\binom{n+m}{m+k}p^k(1-p)^{n-k}.$$

After cancelling p^m, we see that, for $i=0,1,\ldots,n$, the coefficient of $(1-p)^i$ on the left hand side is simply

$$\binom{m+i-1}{i},$$

while on the right hand side the coefficient is

$$\sum_{k=n-i}^{n}\binom{n+m}{m+k}\binom{k}{n-i}(-1)^{k-n+i},$$

and the equality of these two expressions follows from considering the coefficient of w^i on both sides of the identity

$$(1+w)^{m+i-1} = \frac{(1+w)^{n+m}}{(1+w)^{n+1-i}},$$

where w is a dummy variable.

2A.11 Estimating the size of a fish population

A lake contains n fish. A sample of m fish ($m < n$) is caught; the fish are marked, and all are then returned to the lake. Later, a second sample of k fish ($k < n$) is caught. Making reasonable assumptions about the way in which the sampling would be carried out, find an expression for $\Pr(X = x)$, where X is the number of marked fish in the second sample.

This probability is a function of the known quantities k and m, the number x of marked fish in the second sample, and n, which will often be unknown in practice. Viewing $\Pr(X = x)$ as a function $g(n)$ of n, examine the ratio $g(n)/g(n-1)$ and hence obtain the value of n for which $g(n)$ is a maximum.

Solution

In this solution we assume that marking the fish does not affect them in any way, so that marked fish mingle freely with others, and are no more and no less likely to be caught again in the second sample. We assume also that the population is stable, so that there are no births, deaths or migrations during the investigation.

Since k fish are caught in the second sample, this sample can be chosen in $\binom{n}{k}$ different ways. Our assumption about the sampling is that all these ways are equally likely so that each of the possible samples has probability $\binom{n}{k}^{-1}$ of being the one selected.

We now need to count the number of ways of selecting the second sample in which $X = x$. Now if $X = x$, then the number of unmarked fish in the second sample is $k - x$. We can choose the x marked fish in $\binom{m}{x}$ ways, and for each of those ways we can choose the $k - x$ unmarked fish in $\binom{n-m}{k-x}$ ways. Thus the total number of ways of selecting the second sample such that $X = x$ is $\binom{m}{x}\binom{n-m}{k-x}$, each with probability $\binom{n}{k}^{-1}$, and so

$$\Pr(X = x) = \frac{\binom{m}{x}\binom{n-m}{k-x}}{\binom{n}{k}}, \quad \max(0, m+k-n) \le x \le \min(k, m).$$

The bounds shown for x are needed to ensure that the binomial coefficients are valid, and they also correspond to the practical constraints of the problem.

The function $g(n)$ is simply $\Pr(X = x)$ viewed as a function of the unknown n, so we have

$$g(n) = \frac{\binom{m}{x}\binom{n-m}{k-x}}{\binom{n}{k}},$$

and substituting $n - 1$ for n gives

$$g(n-1) = \frac{\binom{m}{x}\binom{n-1-m}{k-x}}{\binom{n-1}{k}},$$

so that

$$\begin{aligned}\frac{g(n)}{g(n-1)} &= \frac{(n-m)!\,(n-k)!\,(n-1)!\,(n-1-m-k+x)!}{(n-m-k+x)!\,n!\,(n-1-k)!\,(n-1-m)!} \\ &= \frac{(n-m)(n-k)}{(n-m-k+x)n} = \eta, \text{ say.}\end{aligned}$$

Now when $\eta > 1$,

$$(n-m)(n-k) > (n-m-k+x)n,$$

i.e.

$$n^2 - nk - mn + mk > n^2 - nm - kn + nx,$$

which results in

$$\frac{mk}{x} > n,$$

as long as $x > 0$. If $x = 0$ then $\eta > 1$ for all values of n, so there is no finite value of n for which $g(n)$ is a maximum. We now turn back to the case $x > 0$, in which case $\eta > 1$ whenever $n < mk/x$. Hence, for $n < mk/x$, $g(n) > g(n-1)$, while, for $n > mk/x$, $g(n) < g(n-1)$. and so the function $g(n)$ takes its maximum value when $n = [mk/x]$, the integer part of mk/x.

Notes

(1) The distribution of X is called the *hypergeometric* distribution, on account of the form of its probability generating function, which is a hypergeometric function. (This distribution also arises in Problem 1A.2.)

In sampling from a finite population, when one counts the number of sample members X with some attribute, the distribution of X is hypergeometric when the sampling is done without replacement (the usual practice). Had the second sample of fish been selected *with* replacement, the distribution of X would have been the binomial distribution $B(k, \frac{m}{n})$, since in this case the k 'trials' are independent, each with the same probability $\frac{m}{n}$ of 'success', i.e. obtaining a marked fish.

When k is small relative to n, the difference between sampling with and without replacement will be slight, and the hypergeometric distribution can be approximated by the binomial distribution above, which is generally easier to manipulate mathematically. Approximation by the normal distribution may also be possible.

(2) If we wish to obtain an estimate of an unknown population size n (and in practice this is often the aim of such an investigation), it is intuitively sensible to select that value of n maximising $\Pr(X = x)$, i.e. $g(n)$. The resulting estimator, $[mk/x]$, is an example of what is termed a *maximum likelihood estimator*. (Other examples can be found in Problems 2A.6 and 4D.4.) As it happens, the estimator mk/x follows from simply equating the proportions of marked fish in the lake and in the second sample, viz.

$$\frac{m}{n} = \frac{x}{k}.$$

For example, suppose that $n = 1000$ (although in practice we would not know this) and that $m = 100$, so that in fact 10% of the fish are marked, and that we select $k = 50$ fish in the second sample. Then we might observe $x = 7$ marked fish. In this event the estimate of n would be 714. In practice, as well as calculating a point estimate for n we would also obtain a confidence interval, but omit this here for reasons of space.

* (3) To illustrate the ideas involved in the method of maximum likelihood, we use a small-scale example. Suppose that in the first sample 5 fish were caught and marked, and that another 3 were caught in the second sample, of which 2 had been marked. We then have $m = 5$, $k = 3$ and $x = 2$, and we wish to estimate n, the total number of fish in the lake. Since this is just an illustration, let us suppose for the moment that there are only two possible values for n, $n = 6$ and $n = 14$. For these two values we can easily calculate $g(n)$, and find that $g(6) = \frac{1}{2}$ and $g(14) = \frac{45}{182} \approx \frac{1}{4}$. Now the function $g(n)$ gives the probability of observing $x = 2$ in the second sample; that is, it gives the *probability of occurrence of what has occurred*. Since $g(6) \approx 2g(14)$, the value $n = 6$ is more plausible than is $n = 14$, since what we have observed to occur is distinctly less likely in the latter case. If we had to estimate n, with only these two values from which to choose, the case for picking $n = 6$ would seem a strong one.

In practice, we are, of course, not restricted to just a couple of possible values for n; the argument above suggests that a sensible way of estimating n is to pick that value for which $g(n)$, the probability of occurrence of what has been observed, is at a maximum. The method is widely used, and the function $g(n)$ is termed the *likelihood function*. The method is then the method of *maximum likelihood*, and the value of n for which the likelihood is maximised is called the *maximum likelihood estimate* of n.

In the case illustrated above, $m = 5$, $k = 3$ and $x = 2$. A little calculation shows that $g(6) = \frac{1}{2}$, $g(7) = \frac{4}{7}$, $g(8) = \frac{15}{28}$, $g(9) = \frac{10}{21}$, and so on. The maximum value is at 7, which is therefore the maximum likelihood estimate of n.

(4) An interesting light can be shed on this method of estimating population sizes by *simulating* the method, using the same known values of n, m and k each time. A histogram could be drawn to summarise the resulting sample of estimates of n, and its location and spread, relative to the (known) value of n can be examined. Typically one finds that the spread can be quite large, unless very large values of m and k are used.

(5) Procedures of this type are known as *capture-recapture experiments*, and are employed in practice, but ways of increasing precision are usually incorporated, such as conducting the sampling more than twice. There are always practical problems to overcome, and the basic assumptions of the method, mentioned above, may not be valid in practice. For example, fish with different histories of capture may have different 'catchability', and one may also have to take into account the fact that the fish population will typically not be stable, as assumed in this problem.

A useful, practical, exercise can result from trying such a sampling experiment in, say, a school playground. It would, at any rate, provide a refreshing alternative to the dreary routine of taking the register twice daily.

2A.12 Collecting cigarette cards

Cigarette packets used to contain cards representing such things as flowers, film stars or football teams. A complete set would comprise n different cards, say, and collectors would aim to obtain a complete set. Clearly, the first packet bought would contain a new card which would start the collection. The next card obtained might simply duplicate the first card, or it might also be a new card, and add to the collection, and so on. Obtain expressions for the mean and variance of the number of packets which must be purchased until a complete set of cards is obtained.

Solution

Let Z denote the total number of packets purchased until the complete set of cards is obtained. Then we can write

$$Z = 1 + X_2 + X_3 + \ldots + X_n,$$

in which the Xs denote independent random variables; X_2 is the number of packets (after the very first) until the second card is added to the collection, X_3 is the number of *extra* packets until the third new card, and so on. Hence

$$E(Z) = 1 + \sum_{i=2}^{n} E(X_i) \tag{1}$$

$$\text{and} \quad \operatorname{Var}(Z) = \sum_{i=2}^{n} \operatorname{Var}(X_i), \quad \text{by independence.} \tag{2}$$

We now obtain the expectation and variance of X_i, the number of packets needed to secure the ith new card, once $i-1$ have been obtained. In this case, for each packet bought, the probability is $(i-1)/n$ that the card it contains is an old one and, correspondingly, the probability is $(n-i+1)/n$ that the card is new. Since, for fixed i, the probability of 'success' (i.e. a packet with a new card) is constant, and the 'trials' are independent, the number of trials until the first success, X_i, has the *geometric* distribution with parameter $(n-i+1)/n$. (This distribution is discussed in more detail in Problem 2A.10.)

We thus see that

$$\Pr(X_2 = k) = \left(\frac{1}{n}\right)^{k-1}\left(\frac{n-1}{n}\right), \quad k = 1, 2, \ldots,$$

$$\Pr(X_3 = k) = \left(\frac{2}{n}\right)^{k-1}\left(\frac{n-2}{n}\right), \quad k = 1, 2, \ldots,$$

and, in general,

$$\Pr(X_i = k) = \left(\frac{i-1}{n}\right)^{k-1}\left(\frac{n-i+1}{n}\right), \quad k = 1, 2, \ldots.$$

Now if a random variable X has a geometric distribution such that

$$\Pr(X = k) = (1-p)^{k-1}p, \quad k = 1, 2, \ldots,$$

then $E(X) = 1/p$ and $\operatorname{Var}(X) = (1-p)/p^2$. Hence we obtain

$$E(X_i) = \frac{n}{n-i+1}$$

and

$$\operatorname{Var}(X_i) = \left(\frac{n}{n-i+1}\right)^2\left(\frac{i-1}{n}\right) = \frac{n(i-1)}{(n-i+1)^2}.$$

Thus, from equations (1) and (2) above,

$$E(Z) = 1 + n\sum_{i=2}^{n}(n-i+1)^{-1}$$

and

$$\operatorname{Var}(Z) = n\sum_{i=2}^{n}(i-1)(n-i+1)^{-2}.$$

These expressions can be written more simply as

$$E(Z) = 1 + n\sum_{j=1}^{n-1} j^{-1}$$

and

$$\operatorname{Var}(Z) = n\sum_{j=1}^{n-1}(n-j)j^{-2}.$$

Notes

(1) As an illustration of these results we consider the case $n = 6$, in which case we obtain the following expectations and variances.

$$E(X_2) = \frac{6}{5}; \quad \operatorname{Var}(X_2) = \frac{6}{25};$$

$$E(X_3) = \frac{3}{2}; \quad \operatorname{Var}(X_3) = \frac{3}{4};$$

$$\mathrm{E}(X_4) = 2\,;\ \mathrm{Var}(X_4) = 2;$$
$$\mathrm{E}(X_5) = 3\,;\ \mathrm{Var}(X_5) = 6;$$
$$\mathrm{E}(X_6) = 6\,;\ \mathrm{Var}(X_6) = 30.$$

Hence the expectation of Z is given by

$$\mathrm{E}(Z) = 1 + 1{\cdot}2 + 1{\cdot}5 + 2 + 3 + 6 = 14{\cdot}7,$$

and its variance is given by

$$\mathrm{Var}(Z) = 0{\cdot}24 + 0{\cdot}75 + 2 + 6 + 30 = 38{\cdot}99.$$

We can see from this information alone that the distribution of Z is likely to be skew, since the minimum possible value, 6, is just 1·4 standard deviations below the mean.

(2) Note that the solution is greatly facilitated by considering Z as the sum of independent random variables. The novelty in the problem lies in the fact that the Xs, while independent, are not identically distributed.

The problem is solved by straightforward, though repeated, use of the geometric distribution, and in particular its mean and variance. These can, naturally, be derived from first principles if necessary, using suitable summations.

(3) When $n = 10$, the distribution of Z forms the basis of what is termed the *coupon-collector test* for randomness of a set of decimal digits.

* (4) Calculation of the mean and variance of Z is, in practice, only the first step in an analysis. One can, of course, go further and derive the distribution of Z, but this is more difficult. We present below an outline of how we proceed in the important case $n = 10$, which (see Note 3) is used to test decimal digits for randomness. We have the equation

$$Z = 1 + \sum_{i=2}^{10} X_i$$

relating Z to the independent random variables $X_2, \ldots, X_{10}$. Because of independence, the probability generating function (or p.g.f.; see Note 4 to Problem 2A.2) is the *product* of the p.g.fs of the component random variables. After some algebra we find that Z has p.g.f. $G_Z(t)$ given by

$$G_Z(t) = \frac{9!\,t^{10}}{\prod_{i=1}^{9}(10 - it)},$$

which may be written in partial fraction form as

$$G_Z(t) = \frac{9t^{10}}{10^8}\sum_{i=1}^{9}(-1)^{9-i}\frac{i^8}{(10 - it)}\binom{8}{i-1}.$$

This can be expanded as a power series in t, and the coefficient of t^j in this series will be the probability that $Z = j$. In this way it can be shown that

$$\Pr(Z = j) = 10^{1-j}\sum_{i=1}^{10}(-1)^{i+1}\binom{9}{i-1}(10 - i)^{j-1},\quad j = 10, 11, 12, \ldots.$$

2B Continuous Distributions

In this section we concentrate principally on the normal distribution, reflecting its central rôle in statistics. A matter of key importance is that of learning how to read tables of the standardised normal distribution, i.e. $N(0,1)$, in the customary notational system in which $N(\mu, \sigma^2)$ represents the normal distribution with mean μ and variance σ^2. Since this is not a conventional textbook we do not include these tables here, but recommend the reader to get to know the layout in several books of statistical tables, so as to appreciate the variety of ways in which the standard information can be presented. Two good sets of tables are those by D. V. Lindley and W. F. Scott (*New Cambridge Elementary Statistical Tables*, Cambridge University Press, 1984) and by H. R. Neave (*Elementary Statistics Tables*, George Allen & Unwin, 1981).

A frequently-used piece of notation in this section, and indeed elsewhere, is $\Phi(z)$, the probability that $Z \leq z$, where $Z \sim N(0,1)$, (read 'Z has the normal distribution with mean 0 and variance 1'). Readers will be familiar with the standardisation property of the normal distribution, that if $X \sim N(\mu, \sigma^2)$, then $(X - \mu)/\sigma \sim N(0,1)$, so that in this case

$$\Pr(X \leq x) = \Phi\left(\frac{x - \mu}{\sigma}\right).$$

2B.1 The slippery pole

A boy is trying to climb a slippery pole and finds that he can climb to a height of at least 1·850 m once in five attempts, and to a height of at least 1·700 m nine times out of ten attempts. Assuming that the heights he can reach in various attempts form a normal distribution, calculate the mean and standard deviation of the distribution. Calculate also the height that the boy can expect to exceed once in one thousand attempts.

Solution

If the boy reaches a height of at least 1·850 m once in five attempts, there is a probability of 0·2 that the height he reaches on any one attempt is at least 1·850 m. Similarly, reaching a height of at least 1·700 m nine times out of ten corresponds to a probability of 0·1 of failing to reach a height of 1·700 m. These probabilities are represented by the shaded areas in Figure 2.1. From tables, the standardised normal deviate corresponding to an upper tail area of 0·2 (i.e. a cumulative probability of 0·8) is 0·8418. Similarly, the standardised normal deviate corresponding to a tail area of 0·1 (cumulative probability of 0·9) is 1·2817.

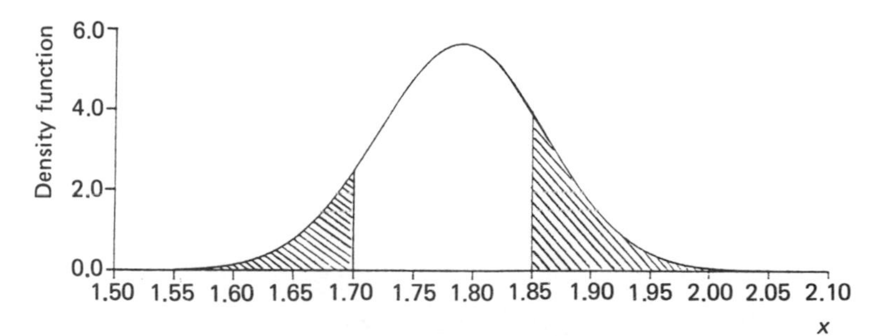

Figure 2.1 P.d.f. of normal distribution for Problem 2B.1

If we denote the unknown mean of the normal distribution by μ and the unknown standard deviation by σ we can write down two equations for μ and σ. These are

$$\frac{1{\cdot}85 - \mu}{\sigma} = 0{\cdot}8418 \qquad (1)$$

and $$\frac{\mu - 1{\cdot}70}{\sigma} = 1{\cdot}2817. \qquad (2)$$

Adding (1) and (2) gives $\frac{0{\cdot}15}{\sigma} = 2{\cdot}1235$. Hence $\sigma = 0{\cdot}0706$ m. Substituting for σ in (1) or (2) gives $\mu = 1{\cdot}7905$ m.

In order to calculate the height the boy can expect to exceed once in one thousand attempts, we find from tables of the normal distribution that the standardised normal deviate corresponding to a tail area of 0·001 is 3·092 (i.e. $\Phi(3{\cdot}092) = 0{\cdot}999$). Hence, if x is the required height,

$$\frac{x - 1{\cdot}7905}{0{\cdot}0706} = 3{\cdot}092,$$

giving $x = 2{\cdot}009$ m.

Notes

(1) An alternative, more concise, way of writing out the derivation of the simultaneous equations (1) and (2) is as follows.

Suppose $\Phi(z)$ is the distribution function of the standard normal distribution $N(0,1)$ and that $\Phi^{-1}(y)$ is the value of z for which $\Phi(z) = y$. Then the information given in the question allows us to write

$$\Phi\left(\frac{1{\cdot}85 - \mu}{\sigma}\right) = 0{\cdot}8$$

and $$\Phi\left(\frac{1{\cdot}70 - \mu}{\sigma}\right) = 0{\cdot}1.$$

Hence $$\frac{1{\cdot}85 - \mu}{\sigma} = \Phi^{-1}(0{\cdot}8) = 0{\cdot}8418$$

and $$\frac{1{\cdot}70 - \mu}{\sigma} = \Phi^{-1}(0{\cdot}1) = -1{\cdot}2817.$$

(2) In problems of this type it is always a good idea to draw a sketch diagram, similar to that shown in Figure 2.1, in order to visualise exactly what is required. (In Figure 2.1, the curve is drawn and calibrated accurately, so that, for example, the shaded region on the right hand side has area exactly 0·2. One can naturally only draw an accurate diagram once one has calculated μ and σ. However, examining the figures given shows that 1·85 must lie above the mean, and 1·70 below it: a sketch diagram showing a normal distribution and the positions of 1·70 and 1·85 in relation to the mean will help to check for gross errors in arithmetic in the solution.)

2B.2 Mixing batches of pesticide

The concentration of a certain active agent in a liquid form of a pesticide must not exceed 12 parts per million. The pesticide is made up in batches in which the concentration varies normally between batches, with mean 8 parts per million and standard deviation 1·5 parts per million. What proportion of batches exceeds the permitted maximum?

If two batches are mixed equally by volume the resulting concentration is the average of the two concentrations in the constituent batches. What proportion of such mixed samples exceeds the permitted maximum? What proportion falls below 3 parts per million (the level at which the pesticide ceases to have any effect)?

Solution

For the first part, the standardised normal deviate of interest is $\dfrac{12-8}{1 \cdot 5} = 2 \cdot 667$. From tables, the corresponding tail area is 0·0038, which is therefore the proportion of batches which exceed the maximum concentration.

If C_1 is the concentration of the first batch and C_2 is the concentration of the second batch (both in parts per million), then the resulting concentration is $C = \frac{1}{2}(C_1 + C_2)$. If the batches are chosen at random, independently of each other, then C has a normal distribution with mean $\frac{1}{2}(8+8) = 8$ and variance $\frac{1}{4}\{(1 \cdot 5)^2 + (1 \cdot 5)^2\} = 1 \cdot 125$, and hence standard deviation 1·0607.

For the proportion which now exceeds the permitted maximum concentration, the standardised normal deviate is $\dfrac{12-8}{1 \cdot 0607} = 3 \cdot 771$, with corresponding tail area $1 - \Phi(3 \cdot 771) = 0 \cdot 0001$, which is the required proportion.

To find the proportion falling below 3 parts per million, we consider the standardised normal deviate $\dfrac{3-8}{1 \cdot 0607} = -4 \cdot 714$, with corresponding tail area $1 - \Phi(4 \cdot 714) = 0 \cdot 0000$, correct to 4 decimal places. (In fact, $\Phi(-4 \cdot 714) \approx 0 \cdot 000\,001\,2$.)

Note

The factor $\frac{1}{2}$ in the expression for the mean of C is easy to remember, but many students find difficulty with the corresponding factor $\frac{1}{4}$ in the variance. This arises because a variance, being the average squared deviation of an observation from its mean, is measured in the square of the original units. Generalising the problem, if we were to mix the batches in unequal proportions w and $1-w$, so that $C = wC_1 + (1-w)C_2$, we would obtain

$$\mathrm{E}(C) = w\,\mathrm{E}(C_1) + (1-w)\mathrm{E}(C_2),$$

but

$$\mathrm{Var}(C) = w^2\,\mathrm{Var}(C_1) + (1-w)^2\,\mathrm{Var}(C_2),$$

C_1 and C_2 being independent.

Note that in no circumstances would one average the standard deviations in this type of problem.

2B.3 Tolerances for metal cylinders

In a manufacturing process circular metal cylinders are being produced as components of a certain product. For a cylinder which is produced to be usable, its length must be between 8·45 cm and 8·65 cm, and its diameter between 1·55 cm and 1·60 cm. The process is running in such a way that the cylinders have lengths which are normally distributed about a mean of 8·54 cm with standard deviation 0·05 cm while, independently, the diameters are normally distributed about a mean of 1·57 cm, with standard deviation 0·01 cm. Find

(i) the percentage of cylinders produced whose lengths fall outside the specified limits;

(ii) the percentage of cylinders produced whose diameters fall outside the specified limits;

(iii) the percentage of cylinders that cannot be used;

(iv) the chance that in a sample of five cylinders taken at random, four are usable and one fails to meet the specifications.

Solution

(i) We need to calculate appropriate tail areas of the distribution of lengths of cylinders, i.e. $N(8{\cdot}54,\ 0{\cdot}05^2)$. The specified limits are 8·45 and 8·65, and the required tail areas are shown as the shaded areas in Figure 2.2.

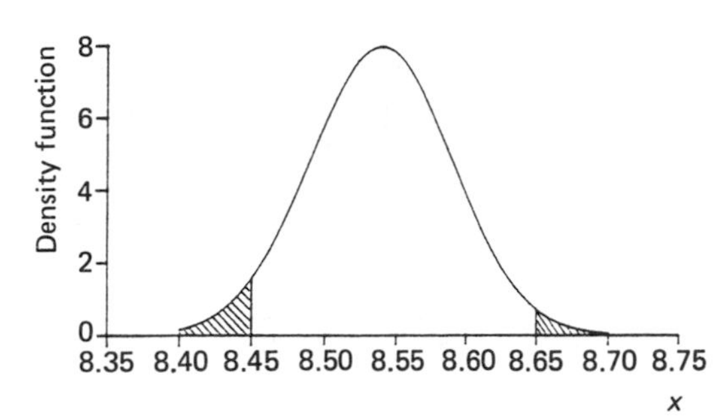

Figure 2.2 Normal distribution with mean 8·54 and s.d. 0·05

To obtain the tail areas we first of all find standardised normal deviates corresponding to the given limits. These are

$$\frac{8{\cdot}65 - 8{\cdot}54}{0{\cdot}05} = 2{\cdot}20,$$

and

$$\frac{8{\cdot}45 - 8{\cdot}54}{0{\cdot}05} = -1{\cdot}80.$$

From tables of the normal distribution, the required areas are $1 - \Phi(2{\cdot}20) = 0{\cdot}0139$ and $\Phi(-1{\cdot}80) = 1 - \Phi(1{\cdot}80) = 0{\cdot}0359$. Therefore the percentage of cylinders whose lengths fail to meet the specifications is $100(0{\cdot}0139 + 0{\cdot}0359) = 4{\cdot}98\%$.

(ii) We now need to find the total area in the tails of the distribution corresponding to standardised normal deviates

$$\frac{1{\cdot}60 - 1{\cdot}57}{0{\cdot}01} = 3$$

and

$$\frac{1{\cdot}55 - 1{\cdot}57}{0{\cdot}01} = -2.$$

From tables, these are $1 - \Phi(3) = 0{\cdot}001\,35$ and $1 - \Phi(2) = 0{\cdot}022\,75$. Hence the percentage of cylinders whose diameters fail to meet the specifications is $100(0{\cdot}001\,35 + 0{\cdot}022\,75)\%$, or $2{\cdot}41\%$.

(iii) Let A be the event that a cylinder chosen at random fails to meet the specified limits for length, and let B be the corresponding event for diameter. Then the event that a cylinder chosen at random from those produced fails to meet the specifications is simply $A \cup B$, and

$$\Pr(A \cup B) = \Pr(A) + \Pr(B) - \Pr(A \cap B).$$

Now, by independence,

$$\Pr(A \cap B) = \Pr(A)\Pr(B) = 0{\cdot}0498 \times 0{\cdot}0241 = 0{\cdot}0012.$$

Hence

$$\Pr(A \cup B) = 0{\cdot}0498 + 0{\cdot}0241 - 0{\cdot}0012 = 0{\cdot}0727,$$

i.e. the percentage of cylinders that will not meet the specifications is $7{\cdot}27\%$.

(iv) Using the binomial distribution with index $n = 5$ and parameter $p = 0{\cdot}9273$ (where p is the probability that a cylinder meets the specifications) we find that the probability that 4 out of 5 cylinders meet the specifications is

$$\binom{5}{4}(0{\cdot}9273)^4(0{\cdot}0727) = 0{\cdot}269.$$

Notes

(1) The first two parts of this problem are standard exercises in computing probabilities from the normal distribution. The third part then uses these in an exercise in combining probabilities of different events, which are not mutually exclusive. Finally, part (iv) uses the probabilities obtained from the normal distribution in an application of the binomial distribution, a common form of problem.

(2) There is a slightly shorter way of solving part (iii) of the problem. Using the notation of that part, the event whose probability is required is the complement of $\overline{A} \cap \overline{B}$. We then obtain, by independence,

$$1 - \Pr(\overline{A} \cap \overline{B}) = 1 - \Pr(\overline{A})\Pr(\overline{B}) = 1-(1-0{\cdot}0498)(1-0{\cdot}0241) = 0{\cdot}0727.$$

2B.4 The number of matches in a box

A manufacturer of matches claims that boxes contain, on average, 49 matches. On checking output over a substantial period, the manager finds that 3% of boxes contain fewer than 46 but 25% contain 51 or more. Making reasonable assumptions, calculate the mean and standard deviation of the number of matches per box. Do you feel that the manufacturer's claim is plausible?

Solution

The problem contains no statement of the *distribution* of the number of matches in a box. The random variable is clearly discrete, but no conventional model seems very plausible. It might be reasonable to assume a fairly symmetrical distribution, and the natural approach to the problem would be to make use of the normal distribution, employing a continuity correction.

Following this approach, and letting μ and σ denote the mean and standard deviation, we have

$$\Phi\left(\frac{46 - \frac{1}{2} - \mu}{\sigma}\right) = 0{\cdot}03$$

$$\text{and} \quad 1 - \Phi\left(\frac{51 - \frac{1}{2} - \mu}{\sigma}\right) = 0{\cdot}25,$$

giving two simultaneous equations in the two unknowns. From normal tables, $\Phi(k_1) = 0{\cdot}03$ gives $k_1 = -1{\cdot}88$, while $\Phi(k_2) = 0{\cdot}75$ gives $k_2 = 0{\cdot}67$. We thus solve

$$\mu - 1{\cdot}88\sigma = 45{\cdot}5$$

$$\text{and} \quad \mu + 0{\cdot}67\sigma = 50{\cdot}5,$$

and these give $\mu = 49{\cdot}19$, $\sigma = 1{\cdot}96$. Since $\mu > 49$, we naturally do not complain about the manufacturer's claim.

Notes

(1) In many examples one *estimates* a population mean or standard deviation. Here the problem is, instead, to *calculate* them. This is because the wording of the problem implies that it is not just sample information that is available. Had the problem referred to, say, results from 200 randomly selected boxes the problem would have been one of inference, and, indeed, rather a difficult one.

(2) The final part of the problem is worded somewhat vaguely, and offers literal-minded mathematicians an opportunity to quibble. Since the calculation does not give μ *exactly* 49 a claim that the average is 49 is not literally correct, but the conclusion drawn in the solution is nonetheless the only reasonable one to draw in the real world.

(3) The comment in the solution about conventional (discrete) models reflects the statistician's natural approach to a problem like the present one. Noting that the random variable is discrete, one runs through a catalogue of plausible models, trying to imagine independent trials with constant probability of success (binomial), number of events in a fixed period (Poisson), and so on. These two are plainly implausible here; in real life the statistician might encounter a problem like this through consultation with the manufacturer, and would then have the opportunity to observe the manufacturing process and use the experience to develop an appropriate model.

(4) A manipulation similar to that used here is considered in greater detail in Problem 2B.1. The two problems differ in principle only in that here we have a discrete random variable, and therefore need to use a continuity correction.

2B.5 Fitting plungers into holes

A random variable X has a $N(\mu_x, \sigma_x^2)$ distribution. Independently, a random variable Y has a $N(\mu_y, \sigma_y^2)$ distribution. What is the distribution of $X-Y$?

In the manufacture of a certain mass-produced article, a circular plunger has to be fitted inside a circular cylinder. It is known that the manufacturing process is adjusted so that the diameters of plunger and cylinder are normally distributed, with mean and standard deviation as follows:

External diameter of plunger: mean 99·7 mm, standard deviation 0·15 mm.
Internal diameter of cylinder: mean 100·2 mm, standard deviation 0·20 mm.

If components are selected at random for assembly, what proportion of plungers will not fit?

Solution

A random variable which is a linear combination of a pair of independently and normally distributed random variables also has a normal distribution, whose mean and variance can be found from the usual rules for sums and differences of independent random variables. Here we note that

$$E(X-Y) = E(X) - E(Y) = \mu_x - \mu_y,$$

and

$$\text{Var}(X-Y) = \text{Var}(X) + \text{Var}(Y) = \sigma_x^2 + \sigma_y^2$$

and therefore find that $X - Y \sim N(\mu_x - \mu_y, \sigma_x^2 + \sigma_y^2)$.

To answer the numerical part we note that a plunger will not fit into a cylinder if the external diameter of the plunger is greater than the internal diameter of the cylinder. Denoting the internal diameter of the cylinder by X and the external diameter of the plunger by Y (both in millimetres), we find that $\mu_x = 100{\cdot}2$, $\mu_y = 99{\cdot}7$, $\sigma_x = 0{\cdot}15$ and $\sigma_y = 0{\cdot}20$, in the notation of the first part of the problem. Hence $X - Y$ has a normal distribution with mean $100{\cdot}2 - 99{\cdot}7 = 0{\cdot}5$ mm, and variance $0{\cdot}15^2 + 0{\cdot}20^2 = 0{\cdot}0625$ mm^2, so that the standard deviation is $0{\cdot}25$ mm.

We now need $\Pr(X - Y < 0)$. To obtain this we use the standardised normal deviate

$$\frac{0 - 0{\cdot}5}{0{\cdot}25} = -2.$$

From tables of the standardised normal distribution, we find the cumulative probability corresponding to this value to be $1 - \Phi(2) = 1 - 0{\cdot}977\,25 = 0{\cdot}022\,75$. This is therefore the proportion of plungers that will not fit inside the cylinders.

Notes

* (1) The first part of the problem requires no more than general results on the expectation and variance of linear combinations of independent random variables, and knowing that linear combinations of normal random variables remain normal. This latter result is quite difficult to demonstrate by direct methods, but is shown easily using the method of *moment generating functions*.

For a random variable X, the moment generating function (or m.g.f.) is defined as $M_X(s) = \mathrm{E}(\mathrm{e}^{sX})$, which can be evaluated as

$$M_X(s) = \int_{-\infty}^{\infty} \mathrm{e}^{sx} f_X(x)\mathrm{d}x,$$

where $f_X(x)$ is the probability density function of X. (The m.g.f. can also be used for a discrete random variable, and an analogous formula is used.) Now the m.g.f. has properties rather similar to those of the probability generating function (see Problem 2A.2) and, in particular, the function can be used to obtain the distribution of the sum of independent random variables.

If we write $W = U + V$, where U and V are independent random variables, then

$$M_W(s) = \mathrm{E}(\mathrm{e}^{sW}) = \mathrm{E}(\mathrm{e}^{sU}\,\mathrm{e}^{sV}) = \mathrm{E}(\mathrm{e}^{sU})\mathrm{E}(\mathrm{e}^{sV}) = M_U(s)M_V(s),$$

where independence of U and V is used to justify writing the expectation of the product $\mathrm{e}^{sU}\,\mathrm{e}^{sV}$ as the product of expectations.

We now use this result to evaluate the distribution of $X - Y$, noting that we require the *difference* between, not the sum of, random variables. We need first to know the form of the moment generating function of $N(\mu, \sigma^2)$: algebraic manipulation gives the result

$$M(s) = \int_{-\infty}^{\infty} \mathrm{e}^{sx} \frac{1}{\sqrt{2\pi}} \mathrm{e}^{-(x-\mu)^2/2\sigma^2}\mathrm{d}x$$

$$= \mathrm{e}^{\mu s + \frac{1}{2}\sigma^2 s^2},$$

where the integral is evaluated by taking the terms in the exponent together and 'completing the square'. We now identify U with X and V with $-Y$; the former is easy, but for the latter we need

$$M_V(s) = \mathrm{E}(\mathrm{e}^{sV}) = \mathrm{E}(\mathrm{e}^{-sY}) = M_Y(-s).$$

We now obtain

$$M_U(s) = M_X(s) = \mathrm{e}^{\mu_x s + \frac{1}{2}\sigma_x^2 s^2}$$

and

$$M_V(s) = M_Y(-s) = \mathrm{e}^{\mu_y(-s) + \frac{1}{2}\sigma_y^2(-s)^2}$$

$$= \mathrm{e}^{-\mu_y s + \frac{1}{2}\sigma_y^2 s^2}.$$

Since $W = U + V$, W has moment generating function

$$M_W(s) = e^{\mu_x s + \frac{1}{2}\sigma_x^2 s^2} e^{-\mu_y s + \frac{1}{2}\sigma_y^2 s^2}$$
$$= e^{(\mu_x - \mu_y)s + \frac{1}{2}(\sigma_x^2 + \sigma_y^2)s^2},$$

and by comparing this expression with the general formula above for the m.g.f. $M(s)$ of $N(\mu, \sigma^2)$ we see that $W \sim N(\mu_x - \mu_y, \sigma_x^2 + \sigma_y^2)$.

(2) In problems like this the concept of 'fitting' can be misleading. What is intended by the problem (which is typical of questions of this nature) is simply whether the plunger can be inserted inside the cylinder, with no implication as to whether the fit is tight or loose, for example. (However, it is not at all difficult to adapt the problem to cope with an extra requirement that the internal diameter of the cylinder should not be more than, say, 1 mm greater than the external diameter of the plunger. The proportion of plungers that fit is then just $\Pr(0 \leq X - Y \leq 1)$ which is easily found to be $\Phi(2) - \Phi(-2) = 0{\cdot}954$.)

(3) It is worth, perhaps, emphasising that the numerical part of the problem requires the use of the result of the theoretical part. This is a common situation, but is often missed by candidates answering examination questions.

2B.6 Bird wingspans

The wingspans of the females of a certain species of bird of prey form a normal distribution with mean 168·75 cm and standard deviation 6·5 cm. The wingspans of the males of the species are normally distributed with mean 162·5 cm and standard deviation 6 cm. What is the probability that, if a male and female are taken at random, the male has a larger wingspan than the female?

Solution

If the wingspan of a female is denoted by X and that of a male by Y then $X - Y$ has a normal distribution with mean $168{\cdot}75 - 162{\cdot}50 = 6{\cdot}25$, and variance $(6{\cdot}5)^2 + 6^2 = 78{\cdot}25$, and hence standard deviation 8·846. We now need to find $\Pr(X - Y < 0)$, the shaded area shown in Figure 2.3. The corresponding value in the standardised normal distribution is

$$\frac{0 - 6{\cdot}25}{8{\cdot}846} = -0{\cdot}707.$$

We require the cumulative probability corresponding to this standardised normal variate; from tables of the normal distribution we obtain

$$1 - \Phi(0{\cdot}707) = 1 - 0{\cdot}760 = 0{\cdot}240.$$

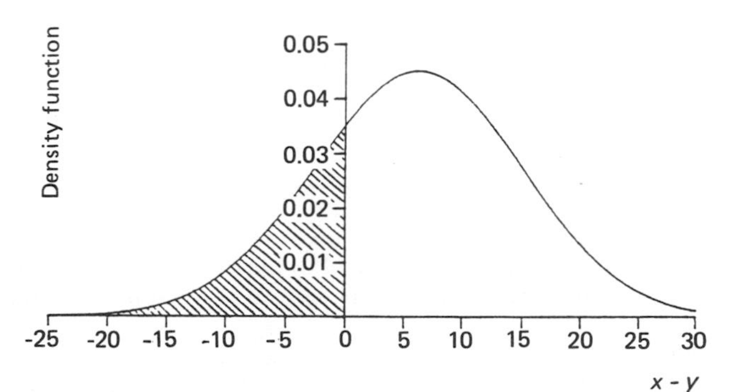

Figure 2.3 Normal distribution with mean 6·25 and variance 78·25

Note

This is a standard problem on the application of the normal distribution. The only point to note is that the problem needs to be tackled by considering the distribution of the difference $X - Y$ between the two random variables; we recall that differences between independent normal random variables are themselves normally distributed, and that the variance of the difference between two *independent* random variables is the sum of their variances. (A more detailed discussion of this topic can be found in Problem 2B.5.)

2B.7 Distributions related to the normal

Random variables $X_1, X_2, \ldots, X_{100}$ are all independent and all have the normal distribution with mean 0 and standard deviation 1. Write down functions of the Xs which have the distributions below, justifying the results you use.

(a) $N(0,4)$

(b) χ^2_{10}

(c) t_{50}

(d) $F_{48,50}$

Solution

(a) $Y_1 = X_1 + X_2 + X_3 + X_4 \sim N(0,4)$.

The mean of Y_1 is the sum of the means, its variance is the sum of the variances (since the Xs are independent) and the sum of normal random variables is normal.

(b) $Y_2 = \sum_{i=1}^{10} X_i^2$ has the χ^2_{10} distribution.

By definition, the square of $N(0,1)$ is χ^2_1, and the sum of n independent χ^2_1 random variables has the χ^2_n distribution.

(c) $Y_3 = \dfrac{X_{51}}{\sqrt{\sum_{i=1}^{50} X_i^2/50}}$ has the t-distribution on 50 degrees of freedom.

By definition, the ratio of an $N(0,1)$ random variable to the square root of an independent χ^2 random variable divided by its number (ν, say) of degrees of freedom has the t-distribution with ν degrees of freedom.

(d) $Y_4 = \dfrac{\sum_{i=51}^{98} X_i^2/48}{\sum_{i=1}^{50} X_i^2/50}$ has the F-distribution with 48 and 50 degrees of freedom.

By definition, the ratio of two independent χ^2 random variables, each divided by the corresponding number of degrees of freedom, has the F-distribution with those numbers of degrees of freedom.

Notes

(1) This problem might be thought unlikely to appear in a public examination, although in fact it is based on a genuine examination question. It does, in any case, serve to illustrate the basic results concerned with sampling distributions related to the normal. These – χ^2, t and F – are crucial to a full understanding of the basis of many common procedures in statistical inference.

For example, squaring Y_3 in part (c) gives a random variable whose numerator X_{51}^2 has the χ_1^2 distribution. Noting the definition in part (d), we see that Y_3^2 is distributed as F with 1 and 50 degrees of freedom. Generalising, we see that the square of a t_ν random variable is $F_{1,\nu}$, a result we shall make use of in a regression problem, Problem 5A.2.

(2) The solution to this problem is, of course, not unique. At a trivial level, renumbering the Xs is obviously acceptable. But there are, as well, several other correct solutions, and we note some of these below.

For part (a), $Y_1^* = 2X_1$ has the required distribution, since the mean remains zero, the variance is increased by a factor of the square of the multiplier and the distribution remains normal. This is a special case of the result that, if $X \sim N(\mu, \sigma^2)$, then $aX + b \sim N(a\mu + b, a^2\sigma^2)$.

For part (b), $Y_2^* = \sum_{i=1}^{11}(X_i - \overline{X})^2$, where $\overline{X} = \sum_{i=1}^{11} X_i/11$, also has the χ_{10}^2 distribution. (This is a natural function to consider because of its link with the sample variance.)

An alternative for part (c) also comes from consideration of normal samples. The most natural t-statistic is

$$Y_3^* = \frac{\overline{X}}{s/\sqrt{n}},$$

where $\overline{X}$ is a sample mean and s the corresponding sample standard deviation; for $n = 51$ we will have 50 degrees of freedom, so set $\overline{X} = \sum_1^{51} X_i/51$ and $s^2 = \sum_1^{51}(X_i - \overline{X})^2/50$.

An alternative to Y_4 comes from an F-test for comparing two sample variances. The numerator and denominator would be of the general form of Y_2^*, each divided by the appropriate number of degrees of freedom. Thus, for example, we could use

$$Y_4^* = \frac{\sum_{i=52}^{100}(X_i - \overline{X}_b)^2/48}{\sum_{i=1}^{51}(X_i - \overline{X}_a)^2/50},$$

where $\overline{X}_a$ is the mean of $X_1, \ldots, X_{51}$ and $\overline{X}_b$ is the mean of the remaining 49 random variables.

2B.8 Failures of belt drives

(a) An exponential distribution has probability density function

$$f(x) = \frac{1}{\alpha}e^{-x/\alpha}, \quad x \geq 0.$$

Show that the mean of the distribution is α and the variance is α^2.

(b) A machine contains two belt drives, of different lengths. These have times to failure which are exponentially distributed, with mean times α and 2α. The machine will stop if either belt fails, and the failures of the belts are independent. Show that the chance that the machine is still operating after a time α from the start is $e^{-3/2}$.

Solution

(a) The mean is given by

$$E(X) = \int_{-\infty}^{\infty} x f(x)dx = \int_0^{\infty} x\frac{1}{\alpha}e^{-x/\alpha}dx.$$

Integrating by parts, this becomes

$$E(X) = \left[-x e^{-x/\alpha}\right]_0^\infty - \int_0^\infty (-e^{-x/\alpha})\, 1\, dx$$

$$= \int_0^\infty e^{-x/\alpha} dx$$

$$= \left[-\alpha e^{-x/\alpha}\right]_0^\infty = \alpha.$$

To obtain the variance we first find

$$E(X^2) = \int_{-\infty}^{\infty} x^2 f(x) dx$$

$$= \int_0^\infty x^2 \frac{1}{\alpha} e^{-x/\alpha} dx.$$

Integrating by parts gives

$$E(X^2) = \left[-x^2 e^{-x/\alpha}\right]_0^\infty - \int_0^\infty (-e^{-x/\alpha})\, 2x\, dx$$

$$= 2\alpha \int_0^\infty x \frac{1}{\alpha} e^{-x/\alpha} dx = 2\alpha E(X) = 2\alpha^2.$$

We now obtain

$$\text{Var}(X) = E(X^2) - \{E(X)\}^2 = 2\alpha^2 - \alpha^2 = \alpha^2,$$

as required.

(b) We require the probability that both belts have failure times greater than α which is, by independence, the product

$$\Pr(\text{first belt has failure time} > \alpha) \times \Pr(\text{second belt has failure time} > \alpha).$$

For the first belt,

$$\Pr(\text{failure time} > \alpha) = \int_\alpha^\infty \frac{1}{\alpha} e^{-x/\alpha} dx$$

$$= \left[-e^{-x/\alpha}\right]_\alpha^\infty = e^{-1}$$

and, for the second belt,

$$\Pr(\text{failure time} > \alpha) = \int_\alpha^\infty \frac{1}{2\alpha} e^{-x/(2\alpha)} dx$$

$$= \left[-e^{-x/(2\alpha)}\right]_\alpha^\infty = e^{-1/2}.$$

Hence the probability that the machine continues to operate after time α is $e^{-1} \times e^{-1/2} = e^{-3/2}$.

Notes

(1) Although the general expressions for $E(X)$, $E(X^2)$ involve integrals evaluated between limits $-\infty$ and ∞, the range of values for which the probability density function is non-zero is more restrictive than this for many distributions. In this example $f(x) = 0$ for $x < 0$; hence the limits for the integrals are 0 and ∞.

* (2) The moment generating function, described in Note 1 to Problem 2B.5 for the normal distribution, can be found very easily for the exponential distribution. By definition

$$M_X(s) = \mathrm{E}(e^{sX}) = \int_{-\infty}^{\infty} e^{sx} f(x)\mathrm{d}x$$

$$= \int_0^{\infty} \frac{1}{\alpha} e^{sx - x/\alpha} \mathrm{d}x$$

$$= \frac{1}{1 - \alpha s}, \quad s < \alpha^{-1}.$$

As could be expected from its name, the moment generating function can be used to obtain the moments of probability distributions. Specifically, the mean is given by $\mathrm{E}(X) = M'_X(0)$, and the second moment $\mathrm{E}(X^2)$ is given by $M''_X(0)$. For the exponential distribution we thus obtain $\mathrm{E}(X) = \alpha$ and $\mathrm{E}(X^2) = 2\alpha^2$, a slightly shorter method than the direct one given in the solution.

2B.9 Guaranteed life of a machine

The lifetime in hours of a certain component of a machine has the continuous probability density function

$$f(x) = \frac{1}{1000} e^{-x/1000}, \quad x \geq 0.$$

The machine contains five similar components, the lifetime of each having the above distribution. The makers are considering offering a guarantee that not more than two of the original components will have to be replaced during the first 1000 hours of use. Find the probability that such a guarantee would be violated, assuming that the components wear out independently, and that if a component does fail then the replacement used is of particularly high quality and will certainly last for the 1000 hours.

Solution

For each component, the probability that the component wears out within the first 1000 hours of use is given by

$$\Pr(\text{lifetime} < 1000 \text{ hours}) = \int_0^{1000} \frac{1}{1000} e^{-x/1000} \mathrm{d}x$$

$$= \Big[- e^{-x/1000} \Big]_0^{1000}$$

$$= 1 - e^{-1} = 0{\cdot}6321.$$

To find the required probability we now use the binomial distribution. Since there are 5 components, acting independently, and all have probability 0·6321 of wearing out, the number X which do wear out is binomially distributed with index (i.e. n) 5 and parameter (i.e. p) 0·6321. We require the probability that X exceeds 2, i.e.

$$\Pr(X = 3) + \Pr(X = 4) + \Pr(X = 5)$$

$$= \frac{5 \times 4}{1 \times 2} (0{\cdot}6321)^3 (0{\cdot}3679)^2 + 5\,(0{\cdot}6321)^4 (0{\cdot}3679) + (0{\cdot}6321)^5$$

$$= 0{\cdot}3418 + 0{\cdot}2937 + 0{\cdot}1009 = 0{\cdot}736.$$

Notes

(1) The first part of the solution is a direct application of the exponential distribution, although to solve the problem one does not have to know this. The exponential distribution is commonly used as a waiting time distribution, or a time-to-failure distribution.

(2) The second part of the solution makes use of the binomial distribution. As in several other problems discussed in this chapter, calculations involving another distribution have to be used first to obtain the value of the binomial parameter p.

(3) Despite the mean lifetime being 1000 hours, the probability that more than two components fail during this period is quite high, at 0·74. It would plainly be very silly of the makers to offer the proposed guarantee!

The reason for the probability being as high as this is simply that the exponential distribution is very skew. Although the mean lifetime is 1000 hours, the probability that an individual lifetime is less than this is as high as 0·6321. Had the distribution been symmetrical, the probability that an individual lifetime would be less than 1000 hours would have been $\frac{1}{2}$, and the distribution of X would thus have been $B(5, \frac{1}{2})$. It is easy to show that for this distribution $\Pr(X > 2) = \frac{1}{2}$, distinctly less than 0·74, although still too high a value to contemplate offering a guarantee.

* (4) Although in the statement of the problem it was made clear (twice!) that failures of replacement components could be ignored, in practice this is an important consideration. The theory underlying the analysis is that of the *Poisson process*, described in Note 1 to Problem 2A.7. In such a process, 'occurrences' (here failures) take place randomly in time, and the total number of these occurrences in a fixed time has a Poisson distribution (and the interval from some starting-point to an occurrence has an exponential distribution, as given above). It can be shown that, for any single original component (and any replacements needed for it) the number of failures in 1000 hours has the Poisson distribution with mean 1; since there are five components, acting independently, the total number of failures Y will have the Poisson distribution with mean 5, so

$$\Pr(Y > 2) = 1 - e^{-5} - 5e^{-5} - \frac{5^2}{2!}e^{-5} = 0{\cdot}875.$$

2B.10 Watch the birdie

A nature reserve has one bird of an uncommon species which frequents the large expanse of reed beds there. It is found that if a birdwatcher arrives at the path adjoining the reed beds, the probability that he or she is still waiting there to see the bird a time y later, where y is measured in minutes, is

$$\frac{1}{3}e^{-\frac{1}{2}y} + \frac{2}{3}e^{-\frac{1}{4}y}, \quad y \geq 0.$$

Find the probability that a birdwatcher spends between 2 and 4 minutes waiting to see the bird, and the mean and variance of the time that he or she waits there.

Solution

Let Y denote the time (in minutes) that the birdwatcher waits on the path. Then we require $\Pr(2 \leq Y \leq 4)$, and are given that

$$\Pr(Y > y) = \frac{1}{3}e^{-\frac{1}{2}y} + \frac{2}{3}e^{-\frac{1}{4}y}.$$

Now

$$\begin{aligned}\Pr(2 \le Y \le 4) &= \Pr(Y \ge 2) - \Pr(Y > 4)\\ &= \left(\frac{1}{3}e^{-1} + \frac{2}{3}e^{-\frac{1}{2}}\right) - \left(\frac{1}{3}e^{-2} + \frac{2}{3}e^{-1}\right)\\ &= 0{\cdot}2366.\end{aligned}$$

To find the mean and variance of Y we use the direct approach based on its probability density function $f_Y(y)$. If the distribution function of Y is $F_Y(y)$, then, for $y \ge 0$,

$$\begin{aligned}F_Y(y) &= 1 - \Pr(Y > y)\\ &= 1 - \left(\frac{1}{3}e^{-\frac{1}{2}y} + \frac{2}{3}e^{-\frac{1}{4}y}\right)\end{aligned}$$

and so the density function $f_Y(y)$ is given by

$$f_Y(y) = \frac{dF_Y(y)}{dy} = \frac{1}{6}\left(e^{-\frac{1}{2}y} + e^{-\frac{1}{4}y}\right).$$

Hence the mean waiting time is given by

$$\begin{aligned}E(Y) &= \int_0^\infty y\frac{1}{6}\left(e^{-\frac{1}{2}y} + e^{-\frac{1}{4}y}\right)dy\\ &= \frac{1}{6}\left[-2ye^{-\frac{1}{2}y} - 4ye^{-\frac{1}{4}y}\right]_0^\infty - \frac{1}{6}\int_0^\infty\left(-2e^{-\frac{1}{2}y} - 4e^{-\frac{1}{4}y}\right)dy\\ &= \frac{1}{3}\left[-2e^{-\frac{1}{2}y} - 8e^{-\frac{1}{4}y}\right]_0^\infty = \frac{10}{3}.\end{aligned}$$

To obtain $\mathrm{Var}(Y)$ we first obtain $E(Y^2)$, given by

$$E(Y^2) = \int_0^\infty y^2\frac{1}{6}\left(e^{-\frac{1}{2}y} + e^{-\frac{1}{4}y}\right)dy$$

We integrate by parts twice, and obtain

$$E(Y^2) = \frac{40}{3} + \frac{8}{3}\left[-4e^{-\frac{1}{4}y}\right]_0^\infty = \frac{72}{3} = 24.$$

The variance of Y is thus given by

$$\mathrm{Var}(Y) = E(Y^2) - \{E(Y)\}^2 = 24 - \frac{100}{9} = \frac{116}{9}.$$

Notes

(1) This problem is a straightforward application of the exponential distribution, also seen in Problems 2B.8 and 2B.9, once it is realised that the expression given in the statement of the problem is not the probability density function. The wording does in fact make this clear, but it is our experience that many students try to answer the problem thinking that what is given is the probability density function.

(2) The problem concerns a *mixture* of exponential distributions, one with mean 2 and one with mean 4 minutes. Such a mixture of distributions has been found very useful in analysing periods of employment, for example, within the civil service, and one often finds questions of this kind set in such a context.

* (3) Sometimes the density function of a random variable can be expressed as a mixture of density functions corresponding to distributions of known mean and variance. In this event, there is an alternative and simpler method, based on the component distributions, of finding the mean and variance. We have here, for the density function of Y,

$$f_Y(y) = \frac{1}{6}\left(e^{-\frac{1}{2}y} + e^{-\frac{1}{4}y}\right), \quad y \geq 0,$$

and can write this as a mixture of exponential density functions, viz.

$$f_Y(y) = \frac{1}{3}\left[\tfrac{1}{2}e^{-\frac{1}{2}y}\right] + \frac{2}{3}\left[\tfrac{1}{4}e^{-\frac{1}{4}y}\right]. \qquad (*)$$

We now make use of the results given in Problem 2B.8 that, for a random variable X with density function $\alpha^{-1}e^{-x/\alpha}$, $E(X) = \alpha$ and $E(X^2) = 2\alpha^2$. Returning to equation (*), and substituting $\alpha = 2$ and $\alpha = 4$ in turn, we obtain

$$E(Y) = \frac{1}{3}(2) + \frac{2}{3}(4) = \frac{10}{3},$$

and

$$E(Y^2) = \frac{1}{3}(2\times 2^2) + \frac{2}{3}(2\times 4^2) = 24,$$

thus giving

$$\begin{aligned} \mathrm{Var}(Y) &= E(Y^2) - \{E(Y)\}^2 \\ &= 24 - \frac{100}{9} = \frac{116}{9}, \quad \text{as before.} \end{aligned}$$

* (4) If Y is a continuous random variable, taking values on the range $(0, \infty)$ only, and with cumulative distribution function $F_Y(y)$, then integration by parts readily reveals that

$$E(Y) = \int_0^\infty \{1 - F_Y(y)\}dy.$$

This result allows one to obtain $E(Y)$ directly from the cumulative distribution function, without needing to construct the density function $f_Y(y)$. Employing this result here, we find

$$\begin{aligned} E(Y) &= \int_0^\infty \Pr(Y > y)dy \\ &= \int_0^\infty \left(\frac{1}{3}e^{-\frac{1}{2}y} + \frac{2}{3}e^{-\frac{1}{4}y}\right)dy \\ &= \frac{2}{3}\int_0^\infty \tfrac{1}{2}e^{-\frac{1}{2}y}dy + \frac{8}{3}\int_0^\infty \tfrac{1}{4}e^{-\frac{1}{4}y}dy \\ &= \left(\frac{2}{3}\times 1\right) + \left(\frac{8}{3}\times 1\right) = \frac{10}{3}, \quad \text{as before.} \end{aligned}$$

* (5) It is important to distinguish between mixtures of density functions and mixtures of random variables. Let X_1 and X_2 be independent random variables with density functions $\frac{1}{2}e^{-\frac{1}{2}x}$, $x \geq 0$, and $\frac{1}{4}e^{-\frac{1}{4}x}$, $x \geq 0$, respectively. Then, from equation (*) in Note 3, Y has a density function which is a mixture of the densities of X_1 and X_2, in proportions $\frac{1}{3}$ and $\frac{2}{3}$ respectively. But if we define a random variable $W = \frac{1}{3}X_1 + \frac{2}{3}X_2$, the density of W is *not* given by equation (*).

The distribution of W is most easily found by using moment generating functions (m.g.fs; see Note 1 to Problem 2B.5 for some details of their definition and use). The m.g.f. $M_W(s)$ of W is defined as

$$M_W(s) = \mathrm{E}(e^{sW}) = \mathrm{E}(e^{\frac{1}{3}sX_1 + \frac{2}{3}sX_2})$$
$$= \mathrm{E}(e^{\frac{1}{3}sX_1})\mathrm{E}(e^{\frac{2}{3}sX_2}),$$

for suitable values of s. But X_1 and X_2 are exponentially distributed random variables, and their m.g.fs take a simple form. We have, for example,

$$\mathrm{E}(e^{\frac{1}{3}sX_1}) = \int_0^\infty e^{\frac{1}{3}sx_1}\tfrac{1}{2}e^{-\frac{1}{2}x_1}\,dx_1$$
$$= \frac{\frac{1}{2}}{\frac{1}{2} - \frac{1}{3}s}, \quad s < \tfrac{3}{2}.$$

Similarly we obtain

$$\mathrm{E}(e^{\frac{2}{3}sX_2}) = \frac{\frac{1}{4}}{\frac{1}{4} - \frac{2}{3}s}, \quad s < \tfrac{3}{8},$$

and so, for $s < \frac{3}{8}$,

$$M_W(s) = \frac{\frac{1}{2}\cdot\frac{1}{4}}{(\frac{1}{2} - \frac{1}{3}s)(\frac{1}{4} - \frac{2}{3}s)}$$
$$= -\tfrac{1}{3}\times\frac{\frac{3}{2}}{\frac{3}{2} - s} + \tfrac{4}{3}\times\frac{\frac{3}{8}}{\frac{3}{8} - s},$$

using partial fractions. But this is the form of a mixture (with weights $-\frac{1}{3}$ and $\frac{4}{3}$) of m.g.fs of exponential distributions (see Note 2 to Problem 2B.8) and we thus obtain

$$f_W(w) = -\tfrac{1}{3}(\tfrac{3}{2}e^{-\frac{3}{2}w}) + \tfrac{4}{3}(\tfrac{3}{8}e^{-\frac{3}{8}w})$$
$$= -\tfrac{1}{2}e^{-\frac{3}{2}w} + \tfrac{1}{2}e^{-\frac{3}{8}w}, \; w \geq 0.$$

2B.11 Breaking a rod at random

(a) A rod of length $2l$ is broken into two parts at a point whose position is random, in the sense that the point is equally likely to be anywhere on the rod. Let X be the length of the smaller part. Write down the probability density function (p.d.f.) of X, and find the expectation of X.

(b) Two rods, each of length $2l$, are independently broken in the manner described above. Let Y be the length of the shortest of the four parts thus obtained. Find the cumulative distribution function $F_Y(y)$ of Y. Hence, or otherwise, show that Y has p.d.f. given by

$$f_Y(y) = \begin{cases} 2(l-y)/l^2, & 0 \leq y \leq l, \\ 0, & \text{otherwise.} \end{cases}$$

Show that the expected value of Y is $l/3$, and find $\mathrm{Var}(Y)$.

(c) For the situation described in part (b), let Z_1 be the sum of the lengths of the two smallest parts and let Z_2 be the sum of the lengths of the smallest and largest of the four parts. Find the mean and variance of Z_1 and Z_2.

Solution

(a) The breakpoint is uniformly distributed along the rod, so that the length X of the smaller part is itself uniformly distributed on the interval $[0, l]$. Its probability density function is thus

$$f_X(x) = \begin{cases} 1/l, & 0 \le x \le l, \\ 0, & \text{otherwise.} \end{cases}$$

The expectation of X can be written down directly as $\frac{1}{2}l$, because of the symmetry of the p.d.f. $f_X(x)$. Alternatively, it can be derived from

$$\begin{aligned} E(X) &= \int_{-\infty}^{\infty} x f_X(x) dx \\ &= \int_0^l \frac{x}{l} dx \\ &= \left[\frac{x^2}{2l}\right]_0^l = \tfrac{1}{2}l. \end{aligned}$$

(b) If X_1 and X_2 are the lengths of the shorter parts of the two rods, then $Y = \min(X_1, X_2)$. We require $F_Y(y)$, the cumulative distribution function of Y, given by

$$F_Y(y) = \Pr(Y \le y) = 1 - \Pr(Y > y),$$

and note that the event $(Y > y)$ will occur if and only if $X_1 > y$ and $X_2 > y$. So

$$\begin{aligned} F_Y(y) &= 1 - \Pr(X_1 > y \cap X_2 > y) \\ &= 1 - \Pr(X_1 > y)\Pr(X_2 > y), \end{aligned}$$

since the two rods are broken independently. Now, for $i = 1, 2$,

$$\begin{aligned} \Pr(X_i > y) &= \int_y^{\infty} f_X(x) dx \\ &= \begin{cases} 1, & y < 0, \\ (l-y)/l, & 0 \le y \le l, \\ 0, & y > l, \end{cases} \end{aligned}$$

and therefore

$$F_Y(y) = \begin{cases} 0, & y < 0, \\ 1 - (l-y)^2/l^2, & 0 \le y \le l, \\ 1, & y > l. \end{cases}$$

Differentiating $F_Y(y)$ with respect to y to obtain the p.d.f. $f_Y(y)$ gives

$$f_Y(y) = \begin{cases} 2(l-y)/l^2, & 0 \le y \le l, \\ 0, & \text{otherwise,} \end{cases}$$

as required.

The expected value of Y is given by

$$\begin{aligned} E(Y) &= \int_{-\infty}^{\infty} y f_Y(y) dy \\ &= \int_0^l 2y(l-y)/l^2 dy, \end{aligned}$$

i.e.

$$E(Y) = \left[\frac{y^2}{l} - \frac{2y^3}{3l^2}\right]_0^l$$

$$= l - 2l/3 = l/3.$$

Now $\mathrm{Var}(Y) = \mathrm{E}(Y^2) - \{\mathrm{E}(Y)\}^2$, and since

$$E(Y^2) = \int_0^l 2y^2(l-y)/l^2 dy$$

$$= \left[\frac{2y^3}{3l} - \frac{y^4}{2l^2}\right]_0^l = \frac{l^2}{6},$$

we obtain

$$\mathrm{Var}(Y) = l^2\left(\frac{1}{6} - \frac{1}{9}\right) = \frac{l^2}{18}.$$

(c) If X_1 and X_2 are as defined in part (b), then $Z_1 = X_1 + X_2$, so

$$E(Z_1) = E(X_1) + E(X_2) = \tfrac{1}{2}l + \tfrac{1}{2}l = l.$$

The rods are broken independently, so that

$$\mathrm{Var}(Z_1) = \mathrm{Var}(X_1) + \mathrm{Var}(X_2)$$

$$= 2\mathrm{Var}(X),$$

since X_1 and X_2 have the same distribution as the random variable X defined in part (a). But

$$E(X^2) = \int_0^l x^2/l \, dx = l^2/3,$$

and so

$$\mathrm{Var}(X) = E(X^2) - \{E(X)\}^2 = l^2/3 - l^2/4 = l^2/12.$$

Hence $\mathrm{Var}(Z_1) = l^2/6$.

To obtain the distribution of Z_2 we note that the shortest and longest parts must come from the same rod, so that necessarily $Z_2 = 2l$, a constant. Hence $E(Z_2) = 2l$ and $\mathrm{Var}(Z_2) = 0$.

Notes

(1) The terms *expected value, expectation* and *mean* are all used in the statement of the problem and in the solution. They do, of course, all have the same meaning.

(2) The results found in the solution to this problem can be used to deduce the mean lengths of the four parts. Let $X_{(i)}$ denote the length of the ith smallest part, for $i = 1, 2, 3$ and 4. Then from part (b) of the solution we find that $E(X_{(1)}) = E(Y) = l/3$, while from part (c) we obtain $E(X_{(1)} + X_{(2)}) = E(Z_1) = l$. We see immediately that $E(X_{(2)}) = l - l/3 = 2l/3$. Further, since each rod was of length $2l$, we have that

$$X_{(1)} + X_{(4)} = X_{(2)} + X_{(3)} = 2l,$$

so that $E(X_{(3)}) = 4l/3$ and $E(X_{(4)}) = 5l/3$. The mean values of the lengths of the four parts are thus in the ratio 1 : 2 : 4 : 5.

2C Simulating Random Variables

The understanding of random variables and probability distributions is often enhanced by simulation, a valuable technique whose power is often underestimated. It has become very widely used since the advent of computers. Almost all computers have a facility for producing random numbers through the RND function in BASIC and equivalent functions in other programming languages. (Note that some versions of BASIC for micro-computers use slightly different forms, for example, RND(1).) A central requirement for serious work is that one must have a reliable source of random digits, and efficient ways of simulating random variables from a variety of probability distributions.

The problems that follow consider simple ways of generating uniform random numbers, and go on to investigate how these may be used to simulate random variables with distributions such as the Poisson and normal.

2C.1 Simulating random variables

Suppose a continuous random variable with a uniform distribution over the range $(0,1)$ is observed to take the value 0·8438. Use this value to generate an observation from each of the following distributions:

(i) the uniform distribution with range $(1,3)$;

(ii) the Poisson distribution with mean 2;

(iii) the normal distribution with mean 10 and variance 25.

Solution

(i) If U has a uniform distribution on the range $(0,1)$, then $X = 1+2U$ has a uniform distribution on the range $(1,3)$; therefore we set $X = 1 + (2\times 0{\cdot}8438) = 2{\cdot}6876$.

(ii) We can construct the cumulative distribution function of a random variable X having a Poisson distribution with mean 2 as follows:

$$\begin{aligned}
\Pr(X=0) &= \mathrm{e}^{-2} = 0{\cdot}1353; \\
\Pr(X=1) &= 2\mathrm{e}^{-2} = 0{\cdot}2707, \quad \text{so } \Pr(X\le 1) = 0{\cdot}4060; \\
\Pr(X=2) &= 4\mathrm{e}^{-2}/2 = 0{\cdot}2707, \quad \text{so } \Pr(X\le 2) = 0{\cdot}6767; \\
\Pr(X=3) &= 8\mathrm{e}^{-2}/3! = 0{\cdot}1804, \quad \text{so } \Pr(X\le 3) = 0{\cdot}8571; \\
\Pr(X=4) &= 16\mathrm{e}^{-2}/4! = 0{\cdot}0902, \quad \text{so } \Pr(X\le 4) = 0{\cdot}9473;
\end{aligned}$$

and so on. We are told that the uniform random variable takes the value 0·8438, and so obtain $X = 3$, since

$$0{\cdot}6767 < 0{\cdot}8438 < 0{\cdot}8571.$$

(See Note 1 for a full justification of this procedure.)

(iii) As in part (ii), we again use the cumulative distribution function, but now of a continuous random variable Z with the standardised normal distribution.

From tables of the standardised normal distribution function, the value z for which $\Pr(Z\le z) = 0{\cdot}8438$ is given by $z = 1{\cdot}010$, which we may use as a value taken by Z. Now if $Z \sim N(0,1)$, then $X = 10+5Z$ has the $N(10,25)$ distribution; hence in order to obtain a value for the random variable X we simply set $X = 10 + (5\times 1{\cdot}010) = 15{\cdot}05$.

Notes

(1) Part (ii) provides a particular example of a general method for simulating discrete random variables. We now describe this method for the case of a non-negative random variable X. (A further example is given in Problem 2C.3.)

If a random variable X has probability function $\Pr(X = i) = p_i$, $i = 0, 1, 2, \ldots$, then we can simulate values of X by first simulating a value for U, a random variable with the uniform distribution on the range $(0, 1)$, and following the rules below:

$$\text{If} \quad U < p_0, \quad \text{set} \quad X = 0; \tag{1}$$

$$\text{if} \quad \sum_{j=0}^{i-1} p_j \le U < \sum_{j=0}^{i} p_j, \quad \text{set} \quad X = i, \quad i = 1, 2, 3, \ldots. \tag{2}$$

We operate these rules by first testing result (1). Then, if $U \ge p_0$, we apply the tests in (2) in sequence, first with $i = 1$, then $i = 2$, and so on, until we find a value of i for which (2) is satisfied.

It is easy to see why the rules enable us to simulate X correctly. For example, we would set $X = 3$ if and only if

$$\sum_{j=0}^{2} p_j \le U < \sum_{j=0}^{3} p_j,$$

and the probability of this event is simply the length of this interval, i.e.

$$(p_0 + p_1 + p_2 + p_3) - (p_0 + p_1 + p_2) = p_3,$$

since U is uniformly distributed over the range $(0, 1)$. So we see that, by implementing the rules, $\Pr(X = 3) = p_3$, as required, and in general $\Pr(X = i) = p_i$.

In many cases this procedure will not be very efficient. (For example, when p_0 is very small then the test: $U < p_0$ will usually fail.) In such cases it is not difficult to devise ways of improving efficiency, and the reader might like to consider how this might be done.

* (2) Once again, in part (iii), we have used a particular example of a general rule, this time for simulating a continuous random variable. To simulate a continuous random variable X with cumulative distribution function $F_X(x)$, we obtain a simulated value u for a random variable U uniformly distributed on the range $(0, 1)$, and simulate X by the value x found by solving the equation

$$F_X(x) = u$$

for x. We now explain why this rule works.

As U is uniformly distributed over $(0, 1)$, $\Pr\{U \le F_X(x)\} = F_X(x)$, and by definition $F_X(x) = \Pr(X \le x)$. Hence

$$\Pr\{U \le F_X(x)\} = \Pr(X \le x).$$

Since $F_X(x)$ is a cumulative distribution function of a continuous random variable it is a monotonic increasing function of x, and so the two events $\{X \le x\}$ and $\{F_X(X) \le F_X(x)\}$ are equivalent and have the same probability. Thus

$$\Pr(X \le x) = \Pr\{F_X(X) \le F_X(x)\},$$

and from the previous displayed equation we see that

$$\Pr\{U \le F_X(x)\} = \Pr\{F_X(X) \le F_X(x)\}.$$

This demonstrates that U and $F_X(X)$ have the same distribution, and so to simulate a value for X we simply obtain a value for U and then solve the equation $F_X(X) = U$ for X, the rule used in the solution.

2C.2 Using dice to obtain random numbers

In a series of ten tosses of two distinguishable dice A and B, the following faces are uppermost (the face for A appearing first in each bracket):

$$(1,3), (3,2), (1,6), (4,2), (2,2), (6,3), (4,6), (5,1), (3,4), (1,4).$$

Explain how you would use the dice to generate uniformly distributed random numbers in the range 0000 to 9999, and illustrate your explanation by obtaining a four-digit random number from the data. Use this random number to obtain a random observation from the binomial distribution with index 3 and parameter 0·3, and a random observation from the exponential distribution with parameter 1.

Solution

There are six possible outcomes for each of A and B, and hence 36 (equally-likely) outcomes in all. If we reject outcomes in which A and B show the same face, i.e. outcomes of the form (i,i), $i = 1,2,\ldots,6$, then we are left with 30 outcomes, which can be used to generate decimal digits as follows.

Outcomes	Digit
(1,2) (1,3) (1,4)	0
(1,5) (1,6) (2,1)	1
(2,3) (2,4) (2,5)	2
(2,6) (3,1) (3,2)	3
(3,4) (3,5) (3,6)	4
(4,1) (4,2) (4,3)	5
(4,5) (4,6) (5,1)	6
(5,2) (5,3) (5,4)	7
(5,6) (6,1) (6,2)	8
(6,3) (6,4) (6,5)	9

Thus, for example, we would choose the digit 5 if we obtain from the dice one of the three outcomes (4,1), (4,2) and (4,3); the probability of this (given that pairs (i,i) are rejected) is

$$\frac{3\times(\frac{1}{6})^2}{\frac{30}{36}} = \frac{1}{10}, \quad \text{as required.}$$

The pairs given thus result in the sequence 0,3,1,5,9,6,6,4,0 of decimal digits, the pair (2,2) being rejected. If we take these digits in fours, viz. 0315, 9664, we obtain the required numbers.

To simulate an observation on a random variable X with the $B(3,0{\cdot}3)$ distribution, we note that

$$\Pr(X=i) = \binom{3}{i}(0{\cdot}3)^i(0{\cdot}7)^{3-i}, \quad i = 0,1,2,3,$$

so that, for example, $\Pr(X=0) = 0{\cdot}343$. Using the first of the four-digit numbers, 0315, and treating it as a uniformly distributed random number on the range (0,1) by preceding it by a decimal point, we see that $0 < 0{\cdot}0315 < 0{\cdot}343$, so we take $X = 0$ as the value for X.

To simulate an exponentially distributed random variable Y with parameter λ, given the value u of a uniformly distributed random variable on the range (0,1), we set

$$Y = -\frac{1}{\lambda}\log_e u.$$

We are given that $\lambda = 1$, and taking $u = 0{\cdot}9664$ as the second four-digit number, 9664, again preceded by a decimal point, we obtain $Y = -\log_e(0{\cdot}9664) = 0{\cdot}0342$.

Notes

(1) An alternative, but more wasteful, way of utilising the dice would be to return the result for the dice as 0 or 1 according as the face shown is one of 1, 2, 3 or one of 4, 5, 6. The resulting binary digits could then be taken four at a time, and treated as binary numbers, so that, for example, 1000 would be 8, 0110 would be 6, and so on. Rejecting values greater than 9 would then result in a sequence of equiprobable digits in the range 0-9, which could then be combined as required.

This approach is wasteful when dice are used, as we have noted. It could, however, be considered if one were using coins instead, by treating 'heads' as 1 and 'tails' as 0.

The fact that we can simulate uniform random numbers in two ways draws attention to the fact that the solution given is not unique – indeed, this is commonly the case in simulation. For example, we could have used *any* way of allocating the 30 results to the 10 digits, as long as just three were assigned to each.

(2) The methods used to simulate the binomial and exponential random variables are as discussed in the notes to Problem 2C.1. To simulate an exponential random variable Y with density function $f_Y(y) = \lambda e^{-\lambda y}$, we can simply set $U = F_Y(Y)$, where U is, as usual, uniformly distributed on the range $(0, 1)$. For the exponential distribution we obtain

$$F_Y(y) = \int_0^y f_Y(x)\mathrm{d}x$$

$$= \int_0^y \lambda e^{-\lambda x}\mathrm{d}x = 1 - e^{-\lambda y},$$

and so we could set $U = 1 - e^{-\lambda Y}$ to give

$$Y = -\frac{1}{\lambda}\log_e(1 - U).$$

But we now note that if U is uniformly distributed on $(0, 1)$, so is $(1 - U)$, and hence we can replace U by $(1 - U)$ in the expression above without affecting the distribution of Y. We can thus use

$$\tilde{Y} = -\frac{1}{\lambda}\log_e U;$$

this is to be preferred, since it involves slightly less arithmetic.

(3) A random variable with the binomial $B(3, 0{\cdot}3)$ distribution can be simulated directly, using the basic definition of such a random variable as the number of successes in 3 independent trials, each with probability 0·3 of success. If, as in the solution, we reject results of the form (i, i), then we could regard the occurrence of any one of the nine outcomes

$$\begin{matrix} (1,2) & (1,3) & (1,4) \\ (1,5) & (1,6) & (2,1) \\ (2,3) & (2,4) & (2,5) \end{matrix}$$

as a success and the occurrence of any other pattern as a failure. Clearly, with this assignment, the probability of a success will be 0·3. For the first three tosses shown in the statement of the problem, viz. (1, 3), (3, 2) and (1, 6), we have, respectively, success, failure and success, so we would record 2 as the first simulated value of the binomial random variable. For the second three results given (excluding the result (2, 2) which is rejected), viz. (4, 2), (6, 3) and (4, 6), we have three failures, so return 0.

This method of simulating Bernoulli trials (whether using dice or computer-generated random numbers) is always available for simulating binomial random variables. It is quite efficient for small values of n, but since the arithmetic (or computer time) required is proportional to n it is less attractive for large values of n.

2C.3 Ants on flower heads

The number of ants on a flower head is a random variable having the following distribution.

Number of ants	0	1	2	3	4	5 or more
Probability	0·2050	0·3162	0·2804	0·1562	0·0422	0

(a) Use the following observations on a random variable with a uniform distribution on $(0, 1)$ to draw a random sample of size four from the distribution of ants:

$$0{\cdot}157\,74,\ 0{\cdot}602\,82,\ 0{\cdot}455\,81,\ 0{\cdot}493\,68.$$

(b) Explain how you would use simulation to estimate the probability that a plant will have 3 or more ants, if the number of flower heads on a plant is a random variable having the Poisson distribution with mean 5.

Solution

(a) If X is a random variable denoting the number of ants on a flower head, then the cumulative distribution function of X is given by the following table.

i	$\Pr(X \le i)$
0	0·2050
1	0·5212
2	0·8016
3	0·9578
4	1·0000

If a random variable U is uniformly distributed on the range $(0, 1)$, then

$$\Pr(0{\cdot}0000 < U \le 0{\cdot}2050) = 0{\cdot}2050 = \Pr(X = 0),$$

$$\Pr(0{\cdot}2050 < U \le 0{\cdot}5212) = 0{\cdot}3162 = \Pr(X = 1),$$

and so on. We can use the general algorithm for simulating discrete random variables (see Note 1 to Problem 2C.1) to simulate X. We see that

$$0{\cdot}0000 < 0{\cdot}157\,74 < 0{\cdot}2050, \quad \text{resulting in } X = 0,$$

$$0{\cdot}5212 < 0{\cdot}602\,82 < 0{\cdot}8016, \quad \text{resulting in } X = 2,$$

$$0{\cdot}2050 < 0{\cdot}455\,81 < 0{\cdot}5212, \quad \text{resulting in } X = 1,$$

$$0{\cdot}2050 < 0{\cdot}493\,68 < 0{\cdot}5212, \quad \text{resulting in } X = 1.$$

The required sample thus consists of the four values 0, 2, 1, 1.

(b) Let Y denote the total number of ants on a randomly selected plant, and let Z denote the number of flower heads on that plant. We are given that

$$\Pr(Z = i) = \frac{e^{-5}5^i}{i!}, \quad i = 0, 1, 2, \ldots.$$

If we write $p = \Pr(Y \ge 3)$, then we need to estimate p using simulation. We can do this by simulating values of Y for n plants (where n is chosen beforehand), and counting the number of times – r, say – for which the simulated value of Y is 3 or more. The estimate of p will then be r/n.

In outline, the simulation proceeds as follows.

(i) Set $r = 0$ and $n_C = 1$.

(ii) Simulate Z, denoting the simulated value by z.

(iii) If $z > 0$, simulate z independent values $x_1, x_2, \ldots, x_z$ for X. (If $z = 0$, then there are no flower heads on the plant, and consequently the plant will not support any ants.)

(iv) If $z > 0$, form the sum $y = \sum_{i=1}^{z} x_i$; if $z = 0$, set $y = 0$.

(v) If $y \geq 3$, increase r by 1.

(vi) Increase n_C by 1.

(vii) If $n_C \leq n$, return to step (ii). Otherwise stop, and record r/n as the estimate of p.

Notes

(1) Part (a) is a straightforward example of the use of a cumulative distribution function to simulate a discrete random variable, given a supply of independent random variables uniformly distributed on the range $(0, 1)$.

(2) The distribution of Y, in part (b), is an example of what is termed a *compound* distribution, since

$$\Pr(Y = i) = \Pr(X_1 + X_2 + \ldots + X_Z = i),$$

where Z itself is a random variable. The distribution of such compound random variables can be found analytically using the technique of probability generating functions. (See Note 4 to Problem 2A.2 for a definition and discussion of properties.) If we define $G_X(t) = \mathrm{E}(t^X)$, $G_Y(t) = \mathrm{E}(t^Y)$ and $G_Z(t) = \mathrm{E}(t^Z)$ as the p.g.fs of X, Y and Z respectively, then the p.g.f. of Y can be shown to be given in general by the result $G_Y(t) = G_Z\{G_X(t)\}$, so that here

$$G_Y(t) = \mathrm{e}^{5\{G_X(t) - 1\}},$$

where by definition

$$G_X(t) = 0{\cdot}2050 + 0{\cdot}3162t + 0{\cdot}2804t^2 + 0{\cdot}1562t^3 + 0{\cdot}0422t^4.$$

We can in fact proceed further with this analysis. If we let $q_i = \Pr(Y \geq i)$, $i = 0, 1, 2, \ldots$, and let $Q(\theta) = \sum_{i=0}^{\infty} \theta^i q_i$, for $0 \leq \theta \leq 1$, then we find

$$Q(\theta) = \sum_{i=0}^{\infty} \theta^i \Pr(Y \geq i) = \sum_{i=0}^{\infty} \theta^i \sum_{j=i}^{\infty} \Pr(Y = j).$$

Reversing the order of summation gives

$$Q(\theta) = \sum_{j=0}^{\infty} \Pr(Y = j) \sum_{i=0}^{j} \theta^i$$
$$= \sum_{j=0}^{\infty} \frac{1 - \theta^{j+1}}{1 - \theta} \Pr(Y = j),$$

since $0 \leq \theta \leq 1$. We thus obtain

$$Q(\theta) = (1 - \theta)^{-1}\{1 - \theta G_Y(\theta)\}.$$

Consequently, $p = \Pr(Y \geq 3)$ is given by the coefficient of θ^3 in the power series expansion of

$$Q(\theta) = (1 - \theta)^{-1}(1 - \theta \mathrm{e}^{5\{G_X(\theta) - 1\}}).$$

After some algebra, we find in the present case that $\Pr(Y \geq 3) = 0{\cdot}902$.

(3) An analytic solution to a problem, such as that found in Note 2 above, is preferable to that found by simulation, although, as in the present problem, the latter may well be easier. If, in the n simulations, a value of Y of 3 or more is obtained R times, then R is a binomially distributed random variable, viz. $R \sim B(n, p)$, so that

$$\text{Var}(R/n) = p(1-p)/n \leq 1/(4n).$$

So, if we require the accuracy of estimation of p from the simulation to be such that, say, $\text{Var}(R/n) \leq \epsilon$, it suffices to choose n such that $4n \geq \epsilon^{-1}$, i.e. to take $n \geq 1/(4\epsilon)$, which could result in a very large value for n. The simulation could then be very time-consuming, and the solution it provides can, of course, only be an approximate one.

By contrast, an analytic solution is exact, and since it is in algebraic form it can be adapted easily. For example, if in the present problem the mean number of flower heads per plant changes from 5 to 6, the only change in the analytic solution is to alter 5 to 6 in the expression for $Q(\theta)$ above.

(4) Some examples of estimating $\Pr(Y \geq 3)$ by simulation, using the method described in the solution, are given in the table below. For each value of n, three simulations were performed.

n	Estimate of $\Pr(Y \geq 3)$ using n simulations
10	1·0
10	0·9
10	0·7
100	0·91
100	0·94
100	0·88
1000	0·908
1000	0·882
1000	0·886
10000	0·9025
10000	0·9079
10000	0·9033

3 Data Summarisation and Goodness-of-Fit

With this chapter we move away from probability and consider part of the process of statistical inference. When one obtains data from a randomly selected sample, the observations are usually of interest not in themselves, but rather in what they can tell us about the population from which they were randomly selected. The precise technique used (a t-test, or a confidence interval for a binomial parameter, for example) will depend, naturally, on circumstances, but there are some general principles applying to most problems.

The most basic of these is that, before undertaking the main analysis, one should wherever possible examine the raw data, in order to discover whether any assumptions which are crucial to that analysis seem likely to be satisfied. A rule of thumb stemming from this is that *plotting the data* is a good way to start any data analysis. Accordingly we discuss some plotting methods in the first of the two sections of this chapter.

In the second section, we carry the process a stage further, in order to present some rather more sophisticated ways of checking assumptions. In many practical problems, a central assumption will be that the data form a random sample from some named distribution; we call the process of matching up distribution to data the 'fitting' of the distribution, and we are naturally concerned to check the quality, or goodness, of the fit.

We have just one cautionary remark. Although it is good practice always to carry out these preliminary analyses, we recognise that it will not always be feasible to do so in the artificial conditions of, say, a three-hour examination; similarly, for reasons of space we have not usually been able to include plots and goodness-of-fit tests for data analysed in Chapters 4 and 5.

3A Data Summarisation

In this short section we present some problems involving the calculation of summary statistics, for example means and variances, from data. The data may be given in grouped form, or the individual observations may be available, and we consider ways of dealing with both types. We discuss also the corresponding problems of graphical presentation of data.

3A.1 The weights of club members

The members of a sports club, 60 male adults, had their weights recorded, in pounds. The weights are given in the table below.

171	160	144	132	154	160	160	158	148	160	131	153
131	165	139	163	149	149	140	149	150	161	136	144
165	174	153	149	157	169	147	156	149	171	149	154
153	149	147	154	145	158	160	152	156	138	167	142
165	155	140	155	158	147	149	169	148	174	150	144

Construct a cumulative frequency table for these weights, using classes of width 5 lb, starting at 129·5 lb. Hence draw a cumulative frequency graph, and use this to find the median and semi-interquartile range.

Use the grouped frequency table to calculate the mean and standard deviation, and compare them with the values obtained using the original, ungrouped, data.

Solution

The number of values falling into each of the classes is given in the table below, together with the cumulative frequencies.

Weight	Frequency	Cumulative Frequency
129·5-134·5	3	3
134·5-139·5	3	6
139·5-144·5	6	12
144·5-149·5	14	26
149·5-154·5	9	35
154·5-159·5	8	43
159·5-164·5	7	50
164·5-169·5	6	56
169·5-174·5	4	60

To obtain the cumulative frequency graph, each cumulative frequency is plotted against the upper boundary of the corresponding class interval. The points plotted are then connected by straight lines. The cumulative frequency graph is shown in Figure 3.1.

The quartiles, usually denoted by Q_1, Q_2 and Q_3, are the values which divide the total frequency into quarters. The second quartile Q_2 is, of course, the median. To obtain the value of Q_1, using the graph, the point on the vertical axis corresponding to $\frac{1}{4}n$ is first found, where $n = 60$. The point on the graph corresponding to this value is then projected down onto the horizontal axis. This latter value is Q_1. For the data, $Q_1 \approx 146$ lb. The values for Q_2 and Q_3 are found similarly, using $\frac{1}{2}n$ and $\frac{3}{4}n$ respectively; we find that $Q_2 \approx 152$ lb and $Q_3 \approx 161$ lb.

The semi-interquartile range is

$$\tfrac{1}{2}(Q_3 - Q_1) = \tfrac{1}{2}(161 - 146) = 7{\cdot}5 \text{ lb}.$$

To calculate the mean and standard deviation we require the class mid-points i.e. 132, 137, 142, 147, 152, 157, 162, 167, 172. To make the calculations slightly easier we recode these values using $x^* = (x - 152)/5$, where x is the original value and x^* the coded value. The coded mid-points are −4, −3, −2, −1, 0, 1, 2, 3, 4. The mean of the coded data is then

$$\frac{3\times(-4) + 3\times(-3) + \ldots + 6\times3 + 4\times4}{60} = \frac{9}{60} = 0{\cdot}15.$$

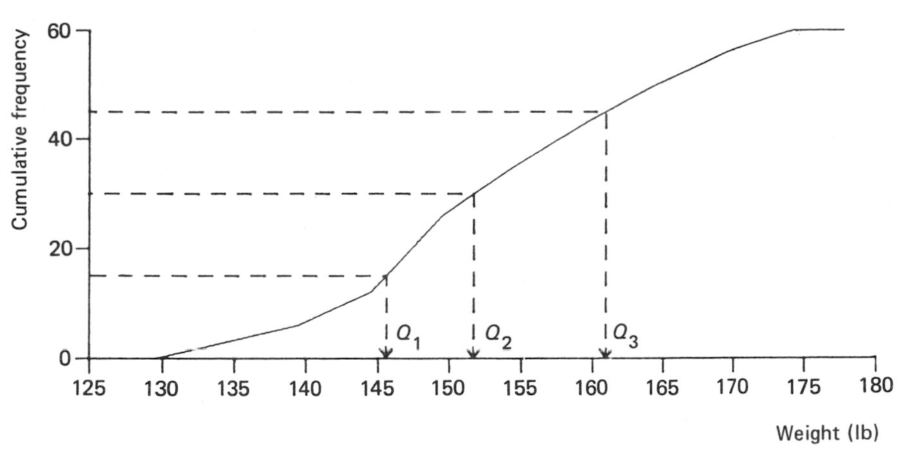

Figure 3.1 Cumulative frequency graph for data on weights

Converting back to the original scale shows the mean to be $152 + 5 \times 0{\cdot}15 = 152{\cdot}75$ lb. The variance of the coded data is

$$s^2 = \frac{1}{60}\left\{3\times(-4)^2 + 3\times(-3)^2 + \ldots + 6\times 3^2 + 4\times 4^2 - \frac{9^2}{60}\right\} = 4{\cdot}4275,$$

where the term 9^2 is the square of the sum of the coded values. The variance in the original scale is then $5^2 \times s^2 = 110{\cdot}69$, in units of pounds, squared. The standard deviation is the square root of this value, i.e. 10·52 lb.

To find the mean and variance of the 60 original observations it is convenient to subtract 100 temporarily from each value. With this informal coding, we find the mean of the raw (i.e. ungrouped) data to be

$$\frac{71 + 60 + \ldots + 50 + 44}{60} = \frac{3175}{60} = 52{\cdot}917,$$

so that the mean weight is 152·917 lb; using the same coding we find that the variance is

$$\frac{1}{60}\left\{71^2 + 60^2 + \ldots + 50^2 + 44^2 - \frac{3175^2}{60}\right\} = 106{\cdot}48.$$

Subtracting 100 does not affect the variance, so that we find the variance of the set of data to be 106·48 and the standard deviation to be its square root, 10·32 lb.

To compare the summary statistics obtained from raw and grouped data, we need only note that the agreement is very good.

Notes

(1) Alternative ways of coding the class mid-points are possible, for example $x^* = (x - 132)/5$. Coding is always optional, and arbitrary; it is done for convenience only, and will not affect the result (once this is transformed to the original scale).

(2) Clearly, if the original data values are available, any summary statistics such as the mean and standard deviation should be calculated from them rather than from a corresponding grouped frequency table. It is always good practice to make use of the full information available.

(3) The frequency distribution is slightly skewed and this is reflected in the mean having a slightly higher value than the median. (A fuller discussion of skewness is given in the Notes to Problem 3A.3.)

(4) Fifty percent of the data lie between the values of Q_1 and Q_3, by definition. If the data given here were normally distributed, then 50% of the data would lie within a range of 0·6745 of a standard deviation on either side of the mean. Since here the standard deviation is 10·32 lb, this range becomes 6·96 lb. This value is slightly smaller than the semi-interquartile range on account of the asymmetry of the distribution.

(5) The values of Q_1, Q_2 and Q_3 could have been obtained directly from the table of cumulative frequencies without drawing the graph. For example, the median may be calculated as

$$Q_2 = L + \frac{n/2 - f}{m} \times c,$$

where, if the class which contains the median value is referred to as the median class, then L is the lower boundary of the median class, m is its frequency, f is the cumulative frequency up to its lower boundary and c is its width.

For the data here we have

$$Q_2 = 149{\cdot}5 + \frac{30 - 26}{9} \times 5 = 151{\cdot}72.$$

The calculation of Q_1 and Q_3 can be done similarly. If n is odd the same formulae are used.

(6) There is no reason why we should restrict our attention to dividing the distribution into *four* equal parts. The nine values which divide the area into ten equal parts are called *deciles* and the ninety-nine values which divide the distribution into one hundred equal parts are called *percentiles*. In general these cut points are called *quantiles*.

(7) In calculating the variance in this problem, the divisor was taken as n, rather than $n-1$. The value $n-1$ is used when the data are considered as a sample from a larger population and the aim is to obtain an estimate of the true (but unknown) population variance. When all that is required is to summarise the data in hand, with no reference to a parent population, the divisor n is generally used.

(8) It is convenient to place here a rather lengthy note on the vexed question of 'n versus $n-1$'. The truth is that in advanced work the divisor to be used is nearly, if not quite, arbitrary; within limits, one can use any divisor. (The discussion in Note 2 of Problem 4D.3 gives an example of an uncommon one.) In a report one must, naturally, say which divisor was used, so that readers can convert to any alternatives they may prefer.

At an elementary level the facts of life are rather different. Statistics is then not an activity, but a subject with a syllabus which is taught and examined. The syllabus may well reflect the view of an examiner, who may well not be professionally involved with statistics and thus may have a partial view, and unfortunately may also have a bee in his bonnet about a 'correct' answer. The painful difficulty is, of course, that what is correct to one examiner may not be to another. Textbooks, likewise, may adopt different conventions.

In this awkward situation the reader will appreciate that we are not going to pretend to produce definitive answers to the vexed question; our case is that no answer could be definitive. But we will attempt to clear some confusion by giving some positive answers, by nailing a few theses to the door, as it were, and by drawing distinctions between similar, but not identical, situations.

The basic problem can be posed in its most confusing way in the form: 'We are given a set x_1, $x_2, \ldots, x_n$ of values. What is the variance of $x_1, x_2, \ldots, x_n$?' Since we will be recommending different answers to the question according to what *sort* of values the xs are, our first thesis is that one should never put the question that way, never think in terms of 'variance of a set of values'. We recall that in probability theory the term 'variance' has a precise definition; when we have a random variable X, its variance $\mathrm{Var}(X)$ is

$$\mathrm{Var}(X) = \mathrm{E}[\{X - \mathrm{E}(X)\}^2].$$

We shall now consider four ways in which these values $x_1, x_2, \ldots, x_n$ can be viewed, and examine the concept of variance for each.

(i) The values have been selected from some larger set or population of values. The interest in the xs lies in what they reveal about the larger set, rather than in themselves.

In these circumstances the xs are clearly a sample, and the almost invariable convention is to use $n-1$ as divisor. To use any other divisor would require corresponding changes to be made to formulae (such as those used in the t-tests) and possibly also to statistical tables. Use of $n-1$ does also have the advantage that the estimator of the variance of the population is unbiased. Despite its arbitrariness, therefore, use of $n-1$ is so common that to fail to use it would be counted a mistake. The resulting quantity is called the *sample variance*; this term almost always connotes a divisor of $n-1$.

(ii) The values form a complete set or population, from which one member is to be chosen at random, with each of the n members having the same chance of being the one chosen.

If we decide to denote by X, say, the value of the population member chosen, then X will be a random variable with a distribution such that

$$\Pr(X = x_i) = \frac{1}{n}, \quad i = 1, 2, \ldots, n.$$

The variance of X can then be obtained from the usual definitional expression

$$\text{Var}(X) = \text{E}[\{X - \text{E}(X)\}^2]$$

and since

$$\text{E}(X) = \frac{1}{n}\sum_{i=1}^{n} x_i = \bar{x}, \text{ say,}$$

it follows that

$$\text{Var}(X) = \frac{1}{n}\sum_{i=1}^{n}(x_i - \bar{x})^2.$$

In this case, where the variance has its usual rôle in relation to a discrete random variable and its probability distribution, the divisor n is indicated.

(iii) The values form a complete set and one wishes to provide a summary measure of their spread, without any implications as to sampling from them.

Here we want to find a way to measure spread, and there is nothing to suggest that concepts based on probability have any relevance. The corrected sum of squares, $\Sigma(x_i - \bar{x})^2$, and the two measures obtained from it by division, are of course worth considering as measures, and the divisor n is generally felt to be better. One reason for this is easiest to see by means of an example. If we have a set of 10 values $x_1, \ldots, x_{10}$, and the corrected sum of squares is 900, then our measure of spread will be 90 if we divide by n, i.e. by 10, or 100 if we divide by 9. Now suppose that the values we have are exactly duplicated, so that our set now consists of 20 values. The corrected sum of squares will now be 1800, and the measure of spread will be found by dividing this either by n, 20, or by $n-1$, 19. Most people would argue that the spread of the 20 values is just the same as for the set of 10. If we use n as divisor, then our measure of spread is $1800/20 = 90$, as before. But using $n-1$ gives $1800/19$, or 94·7, compared with 100 previously.

(iv) The values form a complete set or population, from which a random sample will be selected, without replacement.

There are arguments for both divisors here, but use of $n-1$ does have advantages. In its favour is the fact that, when a random sample is selected without replacement (the usual case when the population is finite) the sample variance will be used to estimate the spread of the population values. The sample variance will conveniently give an unbiased estimator only if the smaller divisor is consistently used, both for sample and population.

It seems then that in the case of a sample there is no real difficulty; one uses the sample variance formula, involving $n-1$. In the case of a population, or complete set, both divisors have their advantages. By and large, those arguments supporting the use of n often prevail in elementary discussions, and that divisor is commonly recommended for the 'variance of a population' in first-level examinations. But the reader will have noticed that we draw this conclusion with some reluctance, since it overlooks more sophisticated arguments and therefore gives a dogmatic prescription which is in fact based on an incomplete assessment.

3A.2 Histogram for catches of fish

A keen angler kept a record of the weight of each of his last 51 catches of fish. The weights, recorded to the nearest 0·1 kg, are as given in the following table.

Weight (kg)	0·0 - 0·4	0·5 - 0·9	1·0 - 1·2	1·3 - 1·7	1·8 - 2·1	2·2 - 3·7	3·8 - 5·2
Frequency	9	12	8	8	8	4	2

Draw a histogram for the data, and use it to calculate the modal class.

Solution

Before the histogram can be drawn, we need to modify the class boundaries given in the problem. This is because the data are continuous, but only recorded to the nearest 0·1 kg. Thus a recorded weight of 0·4 kg represents an actual weight somewhere in the range from 0·35 kg to 0·45 kg, and similarly for the other classes. The only exception is in the first class, where a recorded weight of 0·0 clearly represents a weight in the range (0·0 kg, 0·05 kg).

In a histogram we draw a rectangle for each class so that the area of the rectangle is proportional to the frequency in that class. With equal class widths, the appropriate areas are easily constructed. But when the classes have unequal widths, the heights of the rectangles can no longer be taken equal to the observed frequencies, and rescaling is necessary. For example, classes 0·95-1·25 and 1·25-1·75 both have frequencies of 8, but the class widths are 0·3 and 0·5 respectively. So, if in a histogram the height for the second of these classes is 8, that for the first must be $\frac{0\cdot5}{0\cdot3}\times 8 = 13\frac{1}{3}$.

We now rescale the frequencies in all classes whose class width is not 0·5; by analogy with probability density, we refer to a rescaled frequency as a frequency density. The frequency densities for the classes are given in the table below, and the histogram constructed from the frequency densities is shown in Figure 3.2.

Class boundaries	Frequency density per 0·5 kg
0·00 - 0·45	10·00
0·45 - 0·95	12·00
0·95 - 1·25	13·33
1·25 - 1·75	8·00
1·75 - 2·25	8·00
2·25 - 3·75	1·33
3·75 - 5·25	0·67

From the histogram we see that the modal class is (0·95, 1·25).

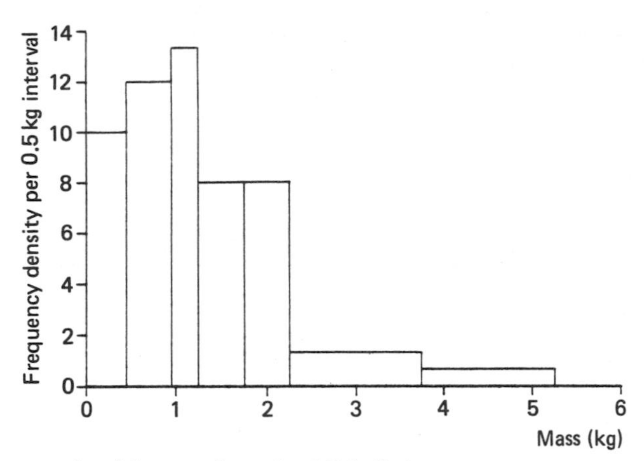

Figure 3.2 Frequency densities: total catch of fish (kg)

Notes

(1) As the weights are recorded to the nearest tenth of a kilogram the class boundaries are taken as 0, 0·45, 0·95, etc. This reflects the continuous nature of the data and the need to ensure that no data value could possibly belong to two different classes. (For a continuous measurement, the chance that an observation is *precisely* 0·45 is negligible, so there is no confusion in specifying this value as a dividing point between two adjacent classes. In problems where measurements are made to a given accuracy, and we are then required to group further, we would have to ensure that no value could appear in two classes. Indeed, in the present problem we could have gone further and specified, for example, the second class as 'above 0·45 but not above 0·95'; but this would have been unnecessarily cumbersome.)

(2) With continuous data, as here, it is usual to quote the *modal class*, that is, 0·95 - 1·25. Sometimes one is expected to calculate a single value for a mode when data are presented in frequency table form, as here. The usual method is a graphical one, and has some theoretical backing when the histogram is constructed with classes of equal width. Of course, in the present case class widths are not all the same, and the method is therefore not applicable.

The basis of the method is that one approximates the histogram, in the region of its maximum, by a quadratic curve. If one were to draw a parabola through three points, the mid-point of the horizontal bar of the histogram for the modal class and corresponding points for the classes either side of the mode, then the maximum of this quadratic could be considered to be an approximation to the mode for the set of data. (A similar justification, valid under the same condition of equal width classes, can be given by constructing a quadratic such that the areas under the curve are equal to the areas under the histogram for the three relevant classes.)

The location of the maximum can be found very simply by drawing two straight lines and finding their intersection. The easiest way to describe the lines is to show how they could be drawn for the present data (although, as we have noted, the method is not applicable when classes are of unequal width). The required lines are always drawn diagonally across the modal class, and for the present data we would join the point (0·95, 12·00) to (1·25, 13·33) and join (0·95, 13·33) to (1·25, 8·00). The intersection of these two lines then indicates the position (although not the height) of the maximum of the quadratic.

3A.3 Distribution of examination marks

The following table shows the number of candidates who scored 0, 1, . . . , 10 marks for a particular question in an examination.

Mark	0	1	2	3	4	5	6	7	8	9	10
No. of Candidates	8	10	49	112	98	86	54	37	28	12	6

Calculate the mean, median and mode of the distribution of marks. What feature of the distribution is suggested by the fact that the mean is greater than the median?

Solution

(a) The mean is the straightforward average of the $8 + 10 + \ldots + 12 + 6 = 500$ marks. Hence the mean is

$$\frac{(8\times0) + (10\times1) + \ldots + (6\times10)}{500} = \frac{2241}{500} = 4{\cdot}48.$$

(b) The median of n values, put in order of magnitude, is the middle value. It is the value ranked in position $(n+1)/2$ if n is odd and is the average of the two middle observations, i.e. those ranked $(n/2)$ and $(n/2 + 1)$, when n is even. In this problem, $n = 500$ and so the median is the average of the 250th and 251st in order of magnitude. These are both 4, so the median is also 4.

(c) The mode is that value which occurs with the greatest frequency, so is clearly 3 here.

(d) The mean, 4·48, is greater than the median, 4, which suggests that the distribution is positively skewed, that is, the values to the right of the median are more spread out than are those on the left. This cannot affect the median, clearly, but the extra spread on the right causes the mean to be greater.

Notes

(1) If a distribution is precisely symmetrical, the mean and median will coincide. By contrast, when the distribution is asymmetrical, and has a relatively long tail to the right (like, for example, the distribution of personal incomes), it is said to be positively skewed. In a similar way a negatively skewed distribution has a relatively long tail to the left. Any skewness is often apparent if a bar chart (or, for continuous data, a histogram) is drawn.

(2) When a distribution is skewed, the median is usually a valuable measure of location. This is because it is unaffected, unlike the mean, by a small number of particularly large (or small) data values, which could be judged not to be of significance in assessing the general location of the distribution. (For example, giving the median income of employed persons in a town will usually tell more about that town's prosperity than giving the mean income.)

The mode is another possible measure of location, but suffers the serious disadvantages that it is not necessarily unique, and is not strictly relevant to sample data from a continuous distribution (but see also Problem 3A.2). In particular, an over-rigorous attitude to samples from continuous distributions might suggest that since the n observations will all be different there will be n modes!

(3) There are several ways of measuring the skewness in a set of sample data. Two which are fairly easy to calculate are

$$\text{(i)}\quad \text{skewness} = \frac{\text{mean} - \text{mode}}{\text{standard deviation}}$$

and

$$\text{(ii)}\quad \text{skewness} = \frac{3(\text{mean} - \text{median})}{\text{standard deviation}}.$$

Both of these make use of the fact that while in a symmetrical distribution with a single mode the mean, median and mode coincide, as asymmetry grows the three gradually drift apart; it is a most rough-and-ready rule that equates the numerators of the expressions above. For the data given in the problem, the standard deviation is 2·00, and the two measures of skewness are, therefore, (i) 0·74 and (ii) 0·72.

However, these two measures are not ideal, as can be seen by examining the data below.

Mark	0	1	2	3	4	5	6	7	8	9	10
No. of Candidates	7	23	88	105	110	114	88	68	36	20	4

It is straightforward to show that the mean, median and mode are 4·58, 4 and 5 respectively. Formula (i) will therefore give a negative value, while formula (ii) gives a positive value. It is apparent from a bar chart of the data that the skewness is positive rather than negative.

Overall, formula (ii) would generally be preferred to (i) because it uses the median rather than the mode; as mentioned in Note 2 the mode may not be unique.

(4) In practice, neither of these two formulae for skewness is often used. Just as the location of the distribution of some random variable X can be measured by $E(X)$ and its spread by the variance $E[\{X-E(X)\}^2]$, the second moment, a measure based on the *third* moment $E[\{X-E(X)\}^3]$ can be used to measure skewness. One can go further and examine higher moments; in particular a measure of ***kurtosis***, the peakedness in the centre of the distribution is based on the fourth moment $E[\{X-E(X)\}^4]$. The measures just mentioned are moments of probability distributions, and the technique extends in a natural way when the population is finite. By analogy with case (ii) in Note 8 to Problem 3A.1 we would, for example, base a measure of the skewness of observations $x_1, x_2, \ldots, x_n$ on the third moment formula $\frac{1}{n}\sum_{i=1}^{n}(x_i - \bar{x})^3$.

These moments are useful in characterising the shape of unusual distributions. They link up, as might be expected, with the moment generating function, discussed briefly in Problem 2B.5; this function, expanded as a power series, can be used to calculate moments conveniently.

3A.4 Consumption of cigarettes

The table below gives data purporting to show the percentages of men and women whose daily consumption of cigarettes lies in various ranges. Draw appropriate pie charts to illustrate the data and bring out the principal features.

Number of cigarettes	Percentage of men	Percentage of women
1-5	3	5
6-10	5	6
11-15	6	7
16-20	7	6
21-30	11	7
31-40	10	5
over 40	8	4

Solution

The production of a pie chart is a fairly trivial matter, and we will not insult the reader by dwelling on the details of arithmetic. The principle is, of course, that the angle assigned to each category, for men and women separately, is in proportion to the number in that category. There is just one point to notice, that the percentages for men add up to 50, those for women to 40 (the remainder being non-smokers), and so the pie charts are constructed so that the total areas are in the corresponding proportions, with the radius of the chart for men being $\sqrt{50}/\sqrt{40}$ of that for women. The charts are presented in Figure 3.3.

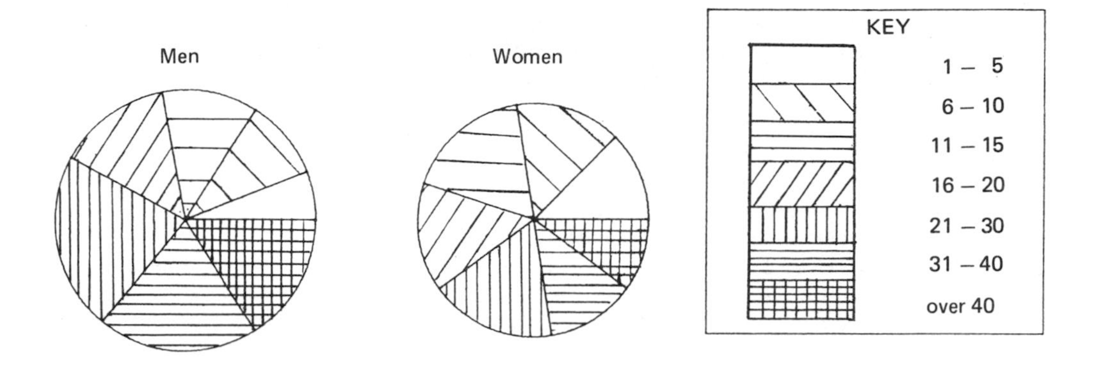

Figure 3.3 Pie charts: numbers of cigarettes smoked daily

The principal feature, shown both by the table and the charts, is that more men than women smoke, and the consumption of cigarettes by male smokers is generally higher than that of women smokers.

Notes

(1) We should emphasise that the figures used above are fictional, chosen so that a few points could be made without excessive calculation.

(2) A decision as to appropriateness of a particular pie chart is by no means as trivial as the production of the chart itself. In the current case there are several alternative ways of presenting the data, each with its advantages. The pie charts shown above are the most basic ones. But one could argue for several alternatives.

(i) The ranges used are unequal, and many statisticians would be happy to sacrifice the detail within the narrower ranges so as to have all the categories of equal size, 1-10, 11-20, and so on. (The final category is awkward, of course, but leaving it as it is would not cause much difficulty.)

(ii) Making the pies of different sizes does bring out the fact that more men than women smoke. But if the aim of the display is to compare individual men and women with other members of their own sexes, as in 'I smoke less than most men, but you smoke more than most women', then equal-sized pie charts would present the information more effectively. (But see also Note 4.)

(iii) A very strong case could be made out for including non-smokers in the pie charts. If this were done, exactly half the pie for men would be assigned to the 'non-smoker' category, with the other categories reduced in proportion; for women similar calculations would be needed. In this case the pies would have equal sizes (unless one wished to be very finicky and take account of the different numbers of males and females in the population).

(3) Yet another pair of pie charts seems possible if we examine the data from the viewpoint of the tobacco companies, wishing to see where the bulk of their trade comes from. Although 7% of women smoke between 11 and 15 cigarettes a day, while only 4% smoke more than 40, more cigarettes in total will be sold for the 4% than for the 7%. The pie charts above do not show this, but the feature can be brought out by looking at the population of cigarettes rather than the population of people.

If we argue that those who smoke between 1 and 5 a day will smoke 3 a day on average, and similarly for the other categories (8, 13, 18, 25·5, 35·5 and (a guess) 45·5), then on average, for every 100 people, the total numbers of cigarettes smoked daily can be found by extending the original table as follows.

Range	Percentage of		Number smoked by	
	men	women	men	women
none	50	60	0	0
1-5	3	5	9	15
6-10	5	6	40	48
11-15	6	7	78	91
16-20	7	6	126	108
21-30	11	7	280·5	178·5
31-40	10	5	355	177·5
over 40	8	4	364	182
Total	100	100	1252·5	800·0

Pie charts could be constructed for the data in the last two columns in the table; note that the chart for men would now have an area larger than that for women by a factor of 1252·5/800·0, since that many more cigarettes are consumed by men than by women. The ratio of the radii would thus be about 1·25, while for the charts in Figure 3.3 the ratio is about 1·12.

(4) Other facets of the data could be brought out by use of other techniques, and indeed it is arguable that since the categories are *ordered,* i.e. form a natural sequence, the use of pie charts will obscure this rather important point. Certainly the production of cumulative frequency polygons or graphs, as in Problem 3A.1, or histograms, as in Problem 3A.2, would be useful.

3B Goodness-of-Fit

Most practical problems in statistical inference have as a basic model that one or more random samples are selected from some distribution or distributions. In some problems the random samples are observed directly, and an analysis such as, for example, a confidence interval for a normal variance, or a two-sample t-test, will be called for. In more complex cases one may have only indirect observations; for example, in regression problems the observations on the dependent variable involve a systematic (linear) component and a random component, and it is the set of random components that are taken to come from some distribution, usually the normal.

The statistical analysis resulting from any assumed model naturally depends on that model: if the model is incorrect then so will be the conclusions, at least in detail. So while it may be acceptable as an academic exercise to assume that observations form a random sample from some distribution, making such an assumption without adequate evidence would be contrary to good scientific practice – and therefore also contrary to good statistical practice.

It follows that we must devise methods of testing assumptions as to distributional shape. There are many such tests, both formal and informal. At simplest, one can construct a histogram and assess by eye whether it is of roughly the required shape; as one would expect, there are more sophisticated methods, discussed in the problems below.

3B.1 Numbers of sixes in dice throws

Two dice were thrown 180 times, and at each throw the number X of sixes was recorded, with the following results.

Number of sixes	(x)	0	1	2	Total
Frequency	(f_x)	105	70	5	180

(a) Test the hypothesis that the distribution of X is binomial with probability p of obtaining a six given by $p = \frac{1}{6}$.

(b) Explain how the test would be modified when the hypothesis to be tested is that X has a binomial distribution, but p is unspecified.

Solution

(a) The appropriate test here is a χ^2 goodness-of-fit test, and the null hypothesis is:

$$H_0 : X \sim B(2, \tfrac{1}{6}).$$

The test statistic C has the form

$$C = \sum_{x=0}^{2} \frac{(f_x - e_x)^2}{e_x},$$

where f_x is the observed frequency of x sixes, and e_x is the corresponding 'expected' frequency when H_0 is true. Now, denoting $\Pr(X = x)$ by p_x, $x = 0, 1, 2$, we obtain, under H_0, $e_x = np_x = 180p_x$, and

$$p_0 = \left(\frac{5}{6}\right)^2, \quad p_1 = 2\left(\frac{5}{6}\right)\left(\frac{1}{6}\right), \quad p_2 = \left(\frac{1}{6}\right)^2,$$

so that $e_0 = 125$, $e_1 = 50$ and $e_2 = 5$. Thus

$$C = \frac{(105-125)^2}{120} + \frac{(70-50)^2}{50} + \frac{(5-5)^2}{5} = 11{\cdot}20.$$

Now, under H_0, the test statistic C has approximately a χ^2 distribution with 2 degrees of freedom, and H_0 will be rejected for large values of C. The upper 1% point of χ_2^2 is 9·21, so we reject H_0 at the 1% significance level.

(b) The null hypothesis is now that X has a binomial distribution, but with parameter p unspecified. The test statistic has the same form as in part (a), except that e_x is now given by $e_x = n\hat{p}_x = 180\hat{p}_x$, where $\hat{p}_x$ is an estimate of the probability that $X = x$, based on the binomial distribution with parameter

$$\hat{p} = \frac{\text{Total number of sixes observed}}{\text{Total number of throws of dice}}.$$

The test statistic C still has, approximately, a χ^2 distribution, but the number of degrees of freedom is reduced by 1.

Notes

(1) The number of degrees of freedom associated with the test statistic is, in general, $k-1$, where k is the number of different classes or values into which the range of the random variable X is divided. However, $k-1$ is only the appropriate number when the distribution of X is fully specified, as in part (a). When parameters have to be estimated, as in part (b), the degrees of freedom are reduced by the number of parameters estimated.

(2) Calculations for the modified test are not specifically asked for in part (b) of the problem, but they are given here for completeness. We have $\hat{p} = 80/360 = 2/9$, so calculate

$$\hat{p}_0 = \left(\frac{7}{9}\right)^2, \quad \hat{p}_1 = 2\left(\frac{7}{9}\right)\left(\frac{2}{9}\right), \quad \hat{p}_2 = \left(\frac{2}{9}\right)^2,$$

Hence $e_0 = 108\frac{8}{9}$, $e_1 = 62\frac{2}{9}$ and $e_2 = 8\frac{8}{9}$, and thus

$$C = \frac{(105 - 108\frac{8}{9})^2}{108\frac{8}{9}} + \frac{(70 - 62\frac{2}{9})^2}{62\frac{2}{9}} + \frac{(5 - 8\frac{8}{9})^2}{8\frac{8}{9}} = 2{\cdot}81.$$

This value of C is to be compared with the χ^2 distribution on 1 degree of freedom, and the null hypothesis is to be rejected if the value lies far enough into the upper tail. The 5% point of the χ_1^2 distribution is 3·84, well above 2·81, so in this case the null hypothesis will not be rejected at any of the conventional significance levels.

(3) The χ^2 goodness-of-fit test generally rejects a null hypothesis only for *large* values of C, since small values imply an exceptionally close fit, with observed frequencies being very close to the expected frequencies. However, if it is suspected that the data are 'fixed' or fraudulent, then the lower tail of χ^2 may also be of interest. Thus, a very small value of C may lead to suspicions that the data had been made up rather than actually observed. (A fairly well known example is the pioneering experiment of Gregor Mendel, on inheritance of colour in sweet pea plants. The observed proportions with particular colours are so close to the theoretical ones that it is generally believed today that the experimenter must have given the data a bit of a helping hand.)

(4) In this problem the notation C is used for the χ^2 goodness-of-fit test statistic, rather than X^2 which is used in most other problems. In the past χ^2 was often used as the symbol for the statistic as well as for the distribution it takes, but this may cause confusion and is rarely done nowadays. The notation X^2 is quite often found, but this is itself not very satisfactory. In the present problem, the binomial random variable was already denoted by the very commonly used X, and use of X^2 as well would have been confusing!

3B.2 The occurrences of thunderstorms

The table below gives the number of thunderstorms reported in a particular summer month by 100 meteorological stations.

No. of thunderstorms	(x)	0	1	2	3	4	5
Number of stations reporting x thunderstorms	(f)	22	37	20	13	6	2

(a) Test whether these data may be reasonably regarded as conforming to a Poisson distribution.

(b) The average number of thunderstorms per month throughout the year is 1·0. Test whether the data above are well fitted by a Poisson distribution with mean 1·0.

(c) The binomial distribution with $n = 5, p = 0{\cdot}3$ provides a good fit to the above data. Without further calculation state why, nevertheless, this binomial model is inappropriate.

Solution

As in Problem 3B.1, the appropriate test for each of parts (a) and (b) is a χ^2 goodness-of-fit test; in the current problem, part (b) has the distribution completely specified, as in part (a) of Problem 3B.1, whereas in part (a) here there is one parameter to be estimated, as in part (b) of Problem 3B.1.

(a) The null hypothesis H_0 is that the random variable X, the number of thunderstorms, has a Poisson distribution with probability function $p(x) = e^{-\mu}\mu^x/x!$, $x = 0, 1, 2, \ldots$, with μ unspecified. The data are given in the form of observed frequencies, f_i, for 6 classes, corresponding to $x = 0, 1, 2, 3, 4$ and 5 or more (see Note 1). The 'expected' frequencies, e_i, required for the goodness-of-fit test statistic are calculated from $e_i = n\hat{p}_i = 100\hat{p}_i$, where $\hat{p}_i$ is obtained from the probability function above with μ replaced by its estimate, the sample mean $\bar{x}$. For the present data, $\bar{x} = 1{\cdot}5$, and

$$e_1 = n\hat{p}_1 = 100e^{-1\cdot5} = 22{\cdot}3,$$

$$e_2 = n\hat{p}_2 = 100\times1{\cdot}5e^{-1\cdot5} = 33{\cdot}5,$$

$$e_3 = n\hat{p}_3 = 100\times\frac{(1{\cdot}5)^2e^{-1\cdot5}}{2!} = 25{\cdot}1,$$

$$\ldots$$

$$e_6 = n - e_1 - e_2 - \ldots - e_5 = 1{\cdot}8.$$

If any of the e_is is smaller than 5, it is usual to combine adjacent classes until all e_is exceed 5 (see Note 3). This will be achieved in the present example if the last two classes are combined, so that there are now 5 classes, and the new final class corresponds to $x \geq 4$.

The χ^2 goodness-of-fit test statistic C can now be computed as

$$C = \sum_{i=1}^{5}\frac{(f_i - e_i)^2}{e_i} = \frac{(22-22{\cdot}3)^2}{22{\cdot}3} + \frac{(37-33{\cdot}5)^2}{33{\cdot}5} + \frac{(20-25{\cdot}1)^2}{25{\cdot}1}$$

$$+ \frac{(13-12{\cdot}6)^2}{12{\cdot}6} + \frac{(8-6{\cdot}5)^2}{6{\cdot}5} = 1{\cdot}76.$$

The test statistic has, approximately, a χ^2 distribution with 3 degrees of freedom (see Note 1 of Problem 3B.1). The computed value is well below the upper 5% point of χ^2_3, which is 7·81, so the Poisson distribution provides a good fit to the present data.

(b) The null hypothesis is now that X has a Poisson distribution with mean $\mu = 1{\cdot}0$. In the χ^2 goodness-of-fit test statistic the f_is are unchanged, but the e_is are calculated from the probability function for the Poisson distribution with mean 1·0, leading to the following values;

$$e_1 = 100\,e^{-1\cdot0} = 36{\cdot}8,$$

$$e_2 = 100e^{-1\cdot0} = 36{\cdot}8,$$

$$e_3 = 100\frac{e^{-1\cdot0}}{2!} = 18{\cdot}4,$$

$$\ldots$$

$$e_6 = 100 - e_1 - e_2 - \ldots - e_5 = 0{\cdot}4.$$

This time it seems appropriate to combine the last three classes so that the new final class corresponds to $x \geq 3$, with expected frequency 8·0. The test statistic is then

$$\sum_{i=1}^{4}\frac{(f_i - e_i)^2}{e_i} = \frac{(22-36{\cdot}8)^2}{36{\cdot}8} + \frac{(37-36{\cdot}8)^2}{36{\cdot}8} + \frac{(20-18{\cdot}4)^2}{18{\cdot}4} + \frac{(21-8{\cdot}0)^2}{8{\cdot}0} = 27{\cdot}22.$$

The test statistic again has, approximately, a χ^2 distribution with 3 degrees of freedom; one degree of freedom was lost compared to part (a) because there is one fewer class, but one was gained because there are no unspecified parameters to estimate. The upper 1% point of χ_3^2 is 11·34, so that a Poisson distribution with mean 1·0 is not a good fit to the data. The physical explanation for this result is probably that thunderstorms are more frequent in the summer than the winter, and that although Poisson distributions may fit individual months' frequencies of thunderstorms, the means of such distributions vary from month to month. Therefore, a common Poisson distribution cannot be fitted to all months of the year.

(c) A binomial distribution with $n = 5$ defines a random variable which can take only the values 0, 1, 2, 3, 4, 5. But X, the number of thunderstorms per month, can clearly exceed 5, so that the binomial distribution cannot be appropriate for X. This illustrates that a good fit of a distribution to a set of data is no guarantee that the distribution is a sensible one.

Notes

(1) The observed frequencies actually correspond to $x = 0, 1, 2, 3, 4$ and 5. However, values of x greater than 5 can occur. (They have non-zero probabilities if a Poisson distribution is assumed.) In a χ^2 goodness-of-fit test, the set of *all* possible values for X is divided into classes, so values of $x \geq 6$ must be included in some class. It is therefore convenient to consider the final observed frequency as corresponding to $x \geq 5$.

(2) Confusion sometimes arises concerning whether 'expected' frequencies, e_i, should be rounded to the nearest integer. The e_is need not be integers, in the same way that expectations of random variables need not be integers for integer-valued random variables (for example, the expected value of the uppermost face when a fair die is thrown is 3·5). In fact, rounding the e_is to the nearest integer will tend to worsen the χ^2 approximation to the distribution of $\sum_{i=1}^{k}(f_i - e_i)^2/e_i$, where k is the number of classes. However, there is no need to calculate the e_is to a high degree of precision; one decimal place is often felt to be sufficient.

(3) The distribution of $\sum_{i=1}^{k}(f_i - e_i)^2/e_i$ is only approximately χ^2, but the approximation is good provided that none of the e_is is too small. The best known 'rule of thumb' is that none of the e_is should be less than 5. If this is not so for the classes as chosen initially, then adjacent classes should be combined until $e_i \geq 5$ for all classes. This procedure has been adopted in the present example, but it is really more stringent than is necessary. In practice, the χ^2 approximation will still be a good one if all e_is are greater than 1, and only a small proportion are less than 5.

Thus, in part (b) of the problem, if only the last two, rather than the last three, classes are combined, then the last two expected frequencies become 6·1 and 1·8. A χ^2 approximation is still reasonable, and there are now 5 classes. The value of the test statistic becomes 35·25, and there are now 4 degrees of freedom. The upper 1% point of χ_4^2 is 13·28, so the conclusion that the Poisson distribution with mean 1 does not provide a good fit is at least as clear-cut as before.

(4) It is not really necessary to look up tables of the χ^2 distribution in part (a). It is useful to remember that the mean of a χ^2 random variable (with ν degrees of freedom, say) is ν, and its variance is 2ν. If the observed value of the test statistic is less than ν, or if it exceeds ν by no more than about $\sqrt{2\nu}$, then it is certainly not necessary to consult χ^2 tables in order to discover that the data provide no significant evidence of lack of fit to the postulated distribution. (This follows because the χ_ν^2 distribution can be approximated by a normal distribution with mean ν and variance 2ν; the approximation is a close one for ν greater than about 100.)

(5) The aim of a goodness-of-fit test is, of course, to determine whether a particular distribution – here the Poisson – can be regarded as a plausible model for the data. Strictly, what we do is to set up and test a null hypothesis that the observations form a random sample from the specified distribution. The natural alternative hypothesis is that the observations form a random sample from some other distribution.

The χ^2 goodness-of-fit test is a general-purpose test, which aims to reject the null hypothesis whenever the data come from another distribution. It is comprehensive, in the sense that, given a large enough sample, it can distinguish between *any* null hypothesis distribution and *any* other distribution. But while this is clearly desirable, it overlooks the fact that, usually, some alternative distributions are much more likely than are others. In the present case, we wish to test the hypothesis that the distribution is Poisson. Now one well-known feature of the Poisson distribution is the equality of mean and variance; it follows that one might consider as a test statistic the sample *index of dispersion*, defined as $I = s^2/\overline{x}$, in a natural notation. The null hypothesis would then be accepted if I were close to 1, and rejected otherwise.

This cannot be regarded as a comprehensive, general-purpose test, since it cannot be used unless the null distribution is Poisson. Nor can it be expected to discriminate between the Poisson and another distribution for which the mean and variance are the same. But in practice a likely alternative to the Poisson distribution is one with a different (usually larger) variance. (See, for example, the discussion of contagious distributions in Note 1 to Problem 2A.7.)

Rather than work directly with the statistic I, we carry out this test in practice by calculating

$$C = (n-1)I = \frac{(n-1)s^2}{\overline{x}} = \frac{\sum_{i=1}^{n} x_i^2 - (\sum_{i=1}^{n} x_i)^2/n}{\overline{x}},$$

where n is the sample size and $x_1, x_2, \ldots, x_n$ are the observations. It turns out that, approximately, $C \sim \chi^2_{n-1}$, and the test is therefore called the χ^2 *index of dispersion* test. Applying the test to the data in the present problem, we find that the sum of the 100 observations is 150 and the sum of their squares is 380; thus

$$\overline{x} = \frac{150}{100} = 1{\cdot}5; \quad (n-1)s^2 = 380 - \frac{150^2}{100} = 155.$$

Hence $C = 155/1{\cdot}5 = 103{\cdot}33$. This is very close to the centre of the χ^2 distribution on 99 degrees of freedom (see Note 4), and we therefore accept the null hypothesis that the data come from a Poisson distribution.

Specialised tests have also been constructed for other distributions (and in particular for the normal distribution). These generally out-perform the χ^2 goodness-of-fit test for the distribution concerned, but cannot sensibly be used for other null hypothesis distributions. (The problem of testing for independence in contingency tables has similar features; we discuss these in Note 1 to Problem 5C.2.)

3B.3 The distribution of I.Q.

Measurements of I.Q. were made for a random sample of 200 grammar school children, with the results given below. Test whether a normal distribution gives a satisfactory fit to the data.

I.Q.	No. of children	I.Q.	No. of children
80-84	1	125-129	8
85-89	3	130-134	2
90-94	16	135-139	2
95-99	33	140-144	1
100-104	44	145-149	2
105-109	31	150-154	0
110-114	26	155-159	2
115-119	20	160-164	1
120-124	8		

Solution

In order to use the χ^2 goodness-of-fit test, 'expected' frequencies are required for all categories. We denote the observed frequency in class i by f_i and the corresponding expected frequency by e_i. To find e_i one must estimate the probability p_i of an observation falling in class i. Then $e_i = n\hat{p}_i$, where $\hat{p}_i$ is the estimate of p_i.

We are fitting a normal distribution, but neither the mean, μ, nor variance σ^2, is given. We therefore estimate μ and σ^2 from the data using $\bar{x}$ and s^2, the sample mean and variance, calculated by assuming that each observation is situated at the midpoint of its class. With this assumption, $\bar{x} = 107{\cdot}58$ and $s^2 = 165{\cdot}38$.

For class i, the value of $\hat{p}_i$ is then obtained, from tables of the standardised normal distribution, by calculating the probability that a normal random variable with mean 107·58 and variance 165·38 takes a value between the lower and upper boundaries of that class. For example, consider the fourth class which has lower and upper boundaries 94·5 and 99·5 respectively. The required probability is

$$\hat{p}_4 = \Phi\left(\frac{99{\cdot}5 - 107{\cdot}58}{\sqrt{165{\cdot}38}}\right) - \Phi\left(\frac{94{\cdot}5 - 107{\cdot}58}{\sqrt{165{\cdot}38}}\right),$$

where the function $\Phi(z)$ is the cumulative distribution function of the standardised normal distribution. Hence

$$\hat{p}_4 = \Phi(-0{\cdot}628) - \Phi(-1{\cdot}017)$$
$$= 0{\cdot}2650 - 0{\cdot}1546 = 0{\cdot}1104,$$

and so $e_4 = 200\hat{p}_4 = 22{\cdot}1$.

This calculation must be repeated for all the other classes (see Note 2), but we find that the e_is for the last six classes are (rounded to one decimal place) 2·3, 0·9, 0·3, 0·1, 0·0 and 0·0 respectively, which are rather small. These classes are therefore combined (see Note 3 to Problem 3B.2), reducing the number of classes to 12. The values of f_i and e_i for these twelve classes are as follows.

Class	f_i	e_i	Class	f_i	e_i
≤ 84	1	7·3	110-114	26	29·1
85-89	3	8·7	115-119	20	23·7
90-94	16	14·9	120-124	8	16·5
95-99	33	22·1	125-129	8	10·0
100-104	44	28·1	130-134	2	5·2
105-109	31	30·8	≥ 135	8	3·6

The χ^2 statistic X^2 is

$$X^2 = \sum_{i=1}^{12} \frac{(f_i - e_i)^2}{e_i},$$

and under the null hypothesis that the data come from a normal distribution it has, approximately, a χ^2 distribution with 9 degrees of freedom (9 = no. of classes − 1 − 2, since two parameters μ and σ^2 were estimated). The value of X^2 is 36·52, which is well above any of the usual percentage points for χ_9^2 ; for example, the 1% point is 21·67. We therefore conclude that the normal distribution does not provide a good fit to these data. (A possible explanation is given in Note 4.)

Notes

(1) There are several points to be made regarding the choice of classes and class boundaries. Frequently the only data given are the numbers of observations in each of a set of pre-determined classes, but sometimes individual observations are given, and classes must be chosen as part of the solution.

When classes are already given, it is usually straightforward to calculate or estimate the probability of falling in each class (under H_0) and hence obtain the e_is. However, there are two possible slight complications.

(i) Class boundaries may not be uniquely defined. This occurs in the present example where the first class contains observations up to 84, and the second class starts at 85. Fitting a normal distribution implies that I.Q. is continuous and could, in theory, take any value between 84 and 85. Assuming that it has been recorded to the nearest whole number, it is natural to take the boundary midway between 84 and 85, at 84·5. However, for other types of data, a different rule might be appropriate for deciding where to put the class boundary. For example, for 'age last birthday' and classes 15-19, 20-24, the boundary would be at 20, rather than at 19·5.

(ii) The first or last class might be open-ended. In the present problem, for example, the first class could have been specified as simply 'I.Q. ≤ 84'. This causes no problems in determining e_i for the class, but it would make the calculation of mean and variance (if needed) ambiguous.

When classes have to be chosen as part of a problem, convenience and common-sense play a large part in the choice, as with the choice of classes for a histogram. For moderate-sized data sets there should be as many classes as possible, subject to none of the e_is being too small. However, there is little point in using more than about twenty classes (only possible with large data sets) unless a very sensitive test is required. Class boundaries should be chosen so that

(a) it is easy to assign observations to classes;

(b) it is easy to look up, or calculate, probabilities, and hence obtain expected frequencies, for each class.

Requirements (a) and (b) may sometimes be difficult to attain simultaneously, so that some trade-off, using common-sense, will be necessary.

(2) For the first class, $\hat{p}_1$ has been calculated as

$$\Phi\left(\frac{84{\cdot}5 - 107{\cdot}58}{\sqrt{165{\cdot}38}}\right),$$

the estimated probability of an I.Q. less than or equal to 84·5. We could have estimated instead the probability of I.Q. falling between 79·5 and 84·5, but we would then have needed to introduce one or more extra classes, all with $f_i = 0$, corresponding to values below 79·5. In theory there is no reason why this should not be done, provided that none of the e_is for the new classes is too small, but in practice it is usual to take the classes as given and to treat the first and last classes as open-ended when calculating e_is.

(3) Since I.Q. tests are generally constructed so that the score has a known mean μ (which is usually 100) and known variance σ^2 for the population as a whole, we might have been required to test the fit of a normal distribution with specified mean and variance. The calculation would proceed as before with μ and σ^2 replacing $\bar{x}$ and s^2, but the test statistic would have two additional degrees of freedom. (A better, more informative, way of arranging these analyses would be first to test the fit of a normal distribution, as done in the solution here; then, if the fit were satisfactory, to assume normality and test hypotheses concerning μ and σ^2. Similar considerations apply to the tests for specific binomial and Poisson distributions discussed in Problems 3B.1 and 3B.2.)

(4) Given the context of the data it is hardly surprising that a normal distribution does not give a good fit. One would expect that there would be some cut-off value below which there were

very few I.Q. values amongst children selected for a grammar school. This would lead to a non-symmetric distribution for I.Q. whose upper tail is much longer than the lower tail, and this is indeed what is observed. Notice that, even without the formal test of significance, the pattern of the differences $(f_i - e_i)$ gives rise to strong suspicions that the normal distribution is not a good fit. The first three differences are negative, followed by four positive differences, then five negative differences, and finally one further positive difference.

3B.4 Frontal breadths of skulls

An anthropologist collected details of 462 skulls of Burmese tribesmen. The data below give the frequencies of occurrence of different values of 'frontal breadths' (measured to the nearest mm) for the skulls. Construct a probability plot for the data, and discuss briefly whether you feel that the data could have been randomly drawn from a normal distribution.

Use your plot to estimate the mean and standard deviation of the distribution.

Range	Frequency	Cumulative Frequency
87-88	5	5
89-90	9	14
91-92	26	40
93-94	34	74
95-96	59	133
97-98	68	201
99-100	80	281
101-102	64	345
103-104	47	392
105-106	42	434
107-108	12	446
109-110	5	451
111-112	3	454
113-114	4	458
115-116	4	462

Solution

In a probability plot the data values are plotted on the arithmetic scale, while the cumulative relative frequencies are plotted on the transformed scale. For grouped data, as here, the data values used are the class boundary points, (88·5, 100·5, . . . , 114·5), and against these are plotted the corresponding cumulative relative frequencies $\frac{5}{462}, \frac{14}{462}, \frac{40}{462}, \ldots, \frac{458}{462}$, all expressed as percentages. (Note that the two extreme points (86·5, 0%) and (116·5, 100%) cannot be plotted.)

The probability plot for the data is shown in Figure 3.4. For clarity we have not drawn a straight line through the points on the plot, but have indicated by small circles two points through which such a line might go.

If the sample (which is, of course, pretty large) were in fact a random sample from a normal distribution, we would expect the probability plot to give virtually a straight line. In fact the plot is, by and large, straight, although the top few points do seem to deviate rather systematically from the line given by the rest. One might reasonably conclude that in the extreme upper tail (about the top 3%) the data depart slightly from the normal shape. But the departure is not great, and would be consistent with two or three observations being recorded

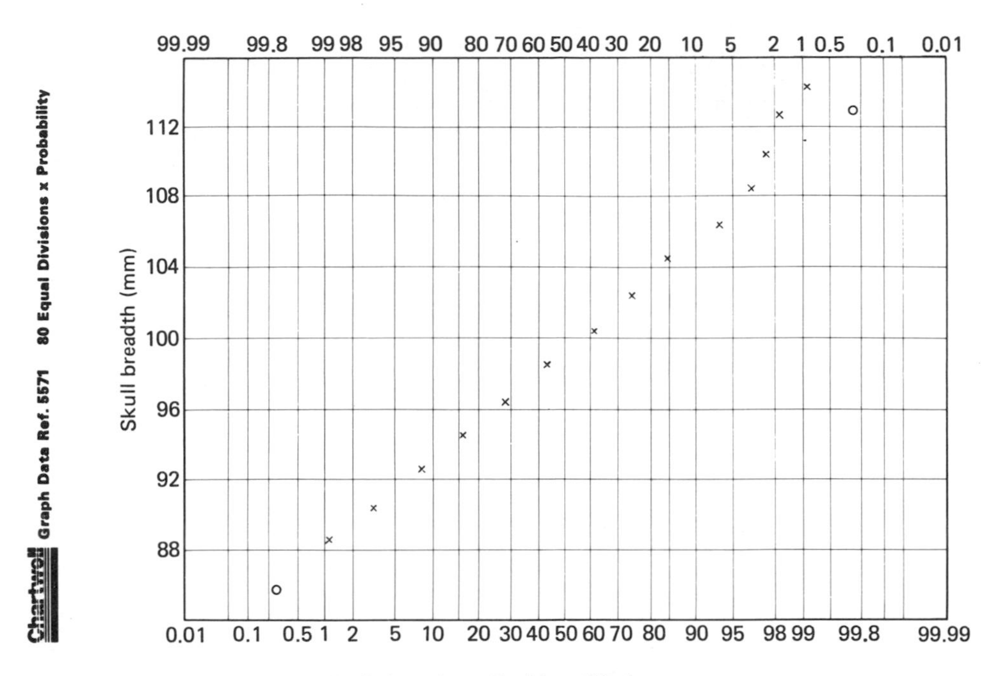

Figure 3.4 Probability plot: skull data from Problem 3B.4
(Based on 'Chartwell' probability paper, produced by H. W. Peel & Company Limited.)

near 115 (exaggeration!) rather than near 110, so in general one would feel that the normal distribution fits the data rather well.

As noted above, we have not superimposed a line on the points in Figure 3.4, but have indicated where a line through most of the points might lie. This 'line' cuts the 50% line at about the point 99·5, and similarly meets the 5% and 95% lines at points 91·6 and 107·4. The difference between these is an estimate of $2\times 1{\cdot}645\sigma$, where σ is the standard deviation of the distribution. We thus find estimates of the mean and standard deviation of the distribution to be 99·5 and 4·80 respectively.

Notes

(1) The basis of the transformation used for the probability scale on probability paper can be expressed very precisely in mathematical terms. For our purposes it may be more helpful to think of it as follows, in terms of an 'infinite' sample, a sample so large that a histogram (with correspondingly narrow class intervals) would take precisely the shape of a normal distribution density function. A probability plot drawn for such an 'infinite' sample would give an exact straight line.

In practice, of course, such a result would be highly suspicious, and one has to decide whether the 'crooked' appearance of a real plot is due to non-normality or to chance fluctuations. As yet, there is no sound, objective, way to do this, but the following consideration may be helpful. As we have seen, an 'infinite' normal sample shows as a straight line. Correspondingly, if one had an 'infinite' sample from some other distribution with a fairly smooth density function, one would expect the probability plot to show as a fairly smooth curve. The question of normality or non-normality thus boils down to assessing whether the plot more nearly resembles points haphazardly scattered about a straight line or about a smooth curve; one must also bear in mind that, with random data, points are necessarily randomly positioned, and must therefore not draw very sweeping conclusions on the basis of the exact positions of one or two points.

(2) In the solution we used the slope and intercept of the probability plot to provide estimates of the parameters μ and σ of the normal distribution. The basis of this method is as follows. If one had a perfect straight line, from an 'infinite' sample, then obviously the 50% point would give the population mean μ, and since 90% of the distribution (from the 5% point to the 95% point) lies in the range $\mu - 1{\cdot}645\sigma$ to $\mu + 1{\cdot}645\sigma$, one could calculate σ. For a real sample, the simplest method of estimating these parameters is to draw a 'good' line through the set of points, and to note the values corresponding to the 5% and 95% points. One then estimates μ by the 50% point, and uses the distance between the other two points as an estimate of $2 \times 1{\cdot}645\sigma$; a simple division by 3·29 thus gives the estimate of σ.

Of course, any two points will suffice to determine the slope of a straight line, so one could use values other than 5% and 95% if it were more convenient, with, naturally, a corresponding change to the divisor. Obviously any such informally produced estimates cannot be expected to be as efficient as the standard $\bar{x}$ and s (in the usual notation), but they do offer speed and simplicity.

(3) The presentation of the data in the statement of the problem is partly helpful and partly distinctly unhelpful. The inclusion of the Cumulative Frequency column is a help, if not a substantial one; typically one would have to calculate this column from the preceding one. The unhelpful part is the first column, which appears to exclude values between 88 and 89, between 90 and 91, etc. The difficulty is resolved by looking above the table, where we find that the data are measured to the nearest millimetre; thus, the class described as 93-94 in fact contains those skulls with breadths between 92·5 and 94·5.

(4) A probability plot is, in essence, a way of testing the goodness of fit of a distribution to some data. The same job could have been done by use of a χ^2 test, as in Problem 3B.3. Conversely, a probability plot could have been done for the I.Q. data of that problem.

3B.5 Rotating bends

A set of 15 observations was made on the number of million cycles to failure of certain industrial components known as 'rotating bends'. It was thought that the logarithms of these quantities might be normally distributed; these logarithms are given below (to base 10, with 1 added for convenience):

0·301	0·519	0·653	0·690	0·892
0·964	0·978	0·987	1·017	1·233
1·342	1·357	1·562	1·845	1·944

Produce a probability plot for the data, and discuss briefly whether you feel that the underlying distribution might be a normal distribution.

Solution

When producing a probability plot for a set of individual observations, one starts by rearranging the sample values in ascending order of magnitude. One then plots the observations, on the linear axis of the probability paper, against appropriate quantities on the probability scale; these quantities are rather like the cumulative relative frequencies of Problem 3B.4, but the underlying ideas are somewhat different. A common practice, which we shall follow here, is to plot observation i $(i = 1, 2, \ldots, n)$ against $i/(n+1)$ (expressed as a percentage). For the current data $n = 15$, so the points plotted are (0·301, 6·25), (0·519, 12·50), . . . , (1·944, 93·75). The resulting plot is shown in Figure 3.5. We find the points to be arranged roughly in a straight line, and conclude that there is no strong suggestion that the data do not come from a normal distribution.

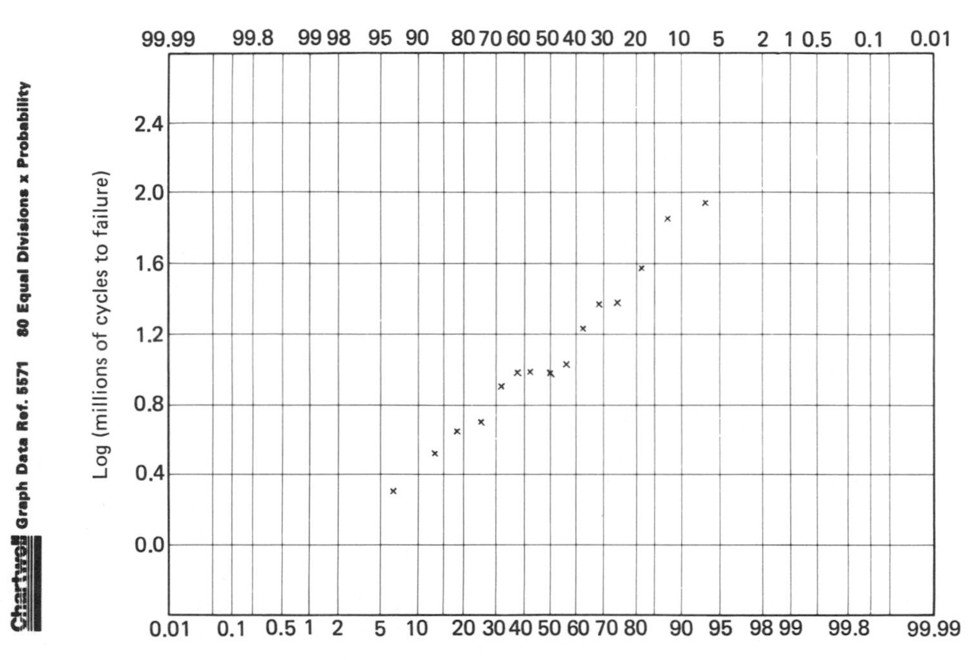

Figure 3.5 Probability plot: rotating bend data of Problem 3B.5
(Based on 'Chartwell' probability paper, produced by H. W. Peel & Company Limited.)

Notes

(1) The technique used here is in essence the same as that for grouped data (see Problem 3B.4), except that we make use of all the information available, and thus plot every point individually. The most natural method might seem to be to plot observation i (in ascending order of magnitude) against i/n (expressed, like others following, as a percentage), since i/n is the proportion of the sample lying at or below the value taken by observation i. This is unsatisfactory, however, for a variety of reasons. (One rather basic reason is that the smallest observation can be plotted (against $1/n$) but not the largest, since n/n, i.e. 100%, is off the edge of the paper.)

A small adjustment overcomes this particular problem; in the solution we used $i/(n+1)$ in place of i/n. This has the advantage of symmetry; the smallest observation (in the current sample) is plotted against $\frac{1}{16}$, or 6·25%, while the largest is plotted against $\frac{15}{16}$, or 93·75% = 100% − 6·25%.

(2) The use of $i/(n+1)$ as the 'plotting position' also has some theoretical backing. This is rather hard to summarise, but we can indicate a key result. Suppose we select a random sample from the uniform distribution on the range $(0,1)$ and plot the values obtained as points on a line of unit length. Clearly, the n points divide the line into $n+1$ regions. The key result is that each of these regions has mean length $(n+1)^{-1}$; in other words, the n points divide the line, on average, into $n+1$ equal parts; so that the average position of observation i is $i/(n+1)$. (While we have presented this result for the special case of the uniform distribution, there is a generalisation covering any continuous distribution, and it is this which gives $i/(n+1)$ its theoretical support.)

(3) For small data sets, the plotting position used for the transformed scale of the probability plot is fairly arbitrary. As noted above, one can object to the position i/n on grounds of lack of symmetry. Such a requirement restricts one, in practice, to plotting positions of the form

$$\frac{i-b}{n+1-2b},$$

but there is no general agreement on the best value for b. Popular choices are $b=0$, used in the solution, and given some justification above, $b=1/2$, used more frequently with grouped data, and in some cases (though we cannot give a justification here) $b=3/8$. The plots from all these positions are, however, virtually indistinguishable, so one's judgement as to the straightness or otherwise of a set of points is not really likely to be affected.

(4) The Notes to Problem 3B.4 are also relevant here. In particular, one can use the probability plot in Figure 3.5 to estimate the parameters of the normal distribution.

(5) The reader will note that the data presented are the logarithms of the number of million cycles to failure of the components, and may wish to take anti-logarithms and thus discover whether the original data could be regarded as normal; this will be an instructive exercise, showing clearly that the normal distribution does not fit the original data.

(6) In many problems in this book, techniques have been used which depend for their validity on the data being normally distributed. In practice, most statisticians would routinely obtain a probability plot, or some equivalent test of normality, before proceeding with such analyses. One must not, of course, expect that with only a few observations available one will be able to decide firmly whether or not the underlying distribution is normal. A random sample of size 15 is still very small (though the plot suggested in Note 5 is decisive enough) and generally one needs at least 20 observations before one can have much confidence that a plot can distinguish normality from non-normality. (This point could usefully be explored with a class, with different members producing probability plots using samples of different sizes.) On the other hand, many techniques discussed later, though assuming normality, are still reasonably effective whenever the distribution is not too obviously non-normal. A probability plot is very useful in reassuring a statistician when this is the case.

3B.6 Distribution of accidents in a factory

The table shows the number of recorded accidents in each week in a large factory, over a period of 100 weeks.

No. of accidents	(x)	0	1	2	3	4	5	6	7 or more
No. of weeks with x accidents	(f)	5	15	21	24	17	11	5	2

Use probability paper

(i) to verify that a Poisson distribution provides a good fit for the data;

(ii) to estimate the mean of the distribution;

(iii) to estimate the probability that a particular future week will have more than 8 accidents, assuming that the same Poisson distribution remains valid in the future.

Solution

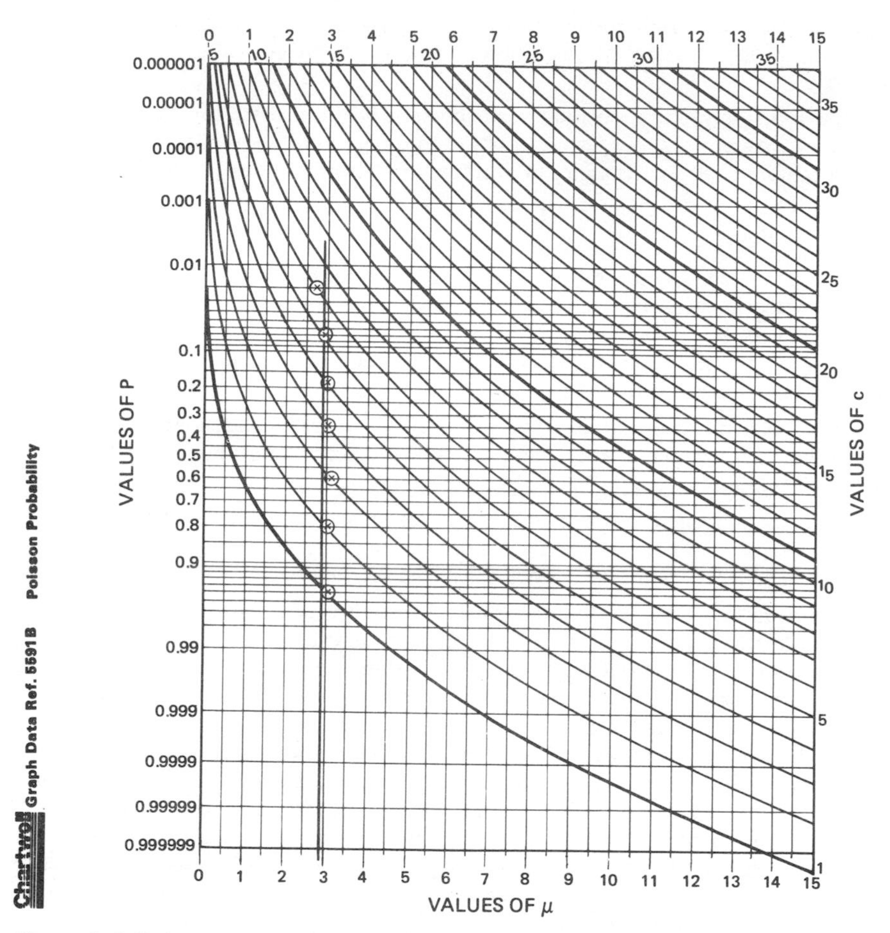

Figure 3.6 Poisson probability plot from accident data (Problem 3B.6)
(Based on 'Chartwell' probability paper, produced by H. W. Peel & Company Limited.)

(i) The data are plotted on Poisson probability paper in Figure 3.6. The assessment of whether a Poisson distribution is a good fit, based on such a plot, is subjective. The idea is that if a Poisson distribution is appropriate then the plotted points will lie close to a straight line parallel to the vertical axis.

The points are plotted on the curved lines which correspond to different values, c, of the random variable X, as indicated on the right-hand scale of the paper. The position of each plotted point is determined by an estimate of $\Pr(X \geq c)$, which is calculated as the proportion of the data taking values of c or more. The left-hand scale on the paper gives this exceedance probability.

In the present example the points do indeed lie close to such a line, with no obvious systematic deviation, so we conclude that the Poisson distribution provides a good fit to the data.

(ii) If exact Poisson probabilities are plotted on Poisson probability paper, then they lie on a vertical line, whose intercept on the horizontal axis is the mean of the distribution. Thus the intercept of the fitted line provides an estimate of the mean. For the present example, the estimate is somewhere between 2·9 and 3·0; it is impossible to be very precise when fitting the line by eye. (Counting the class '7 or more accidents' as '7 accidents', the sample mean number of accidents in a week is 2·96.)

(iii) As stated above, the left-hand scale on Poisson probability paper gives the exceedance probabilities for a Poisson random variable with mean given on the horizontal scale. Observing where the line $c = 9$ intersects the plotted vertical straight line, and reading off on the left hand scale, gives an estimate of the probability of more than eight accidents. The estimate is 0·003.

Note

The present problem illustrates the use of one of the less well-known types of probability paper, that for the Poisson distribution. Because such paper is relatively rare, we have included notes at the appropriate points in the solution, rather than gathering them together.

4 Inference

To most statisticians inference lies at the heart of the subject. The process of inference, of drawing conclusions from sample data about the entire, but unobservable, group from which the sample was randomly drawn, is in essence the universal process of scientific discovery and shares its obvious importance.

Applications of inference are naturally frequent in scientific areas, but not only in those areas. Inference is used in opinion polling and in any form of sample enquiry, in educational research, and indeed in a vast range of applications. In this book we cover many different types of problem in which inferences are required. Some types, for example regression, analysis of variance and goodness-of-fit testing problems, are dealt with in other chapters, but it is important to keep in mind that all are particular facets of inference.

In this chapter we concentrate on relatively straightforward cases involving the binomial, Poisson and normal distributions. But even with this restriction the scope is still wide. Thus we may be working with a single random sample, or may be comparing results from two samples. We may (in the case of the normal distribution, particularly) be working with an extra, unknown, parameter, or may have no such difficulty. The inferences may be required either in the form of confidence intervals or tests of hypotheses, and in the latter case we may need one-tailed or two-tailed tests.

We have divided this chapter into four sections. The first two deal with the normal distribution, Section 4A deals with problems involving a single sample and Section 4B with two-sample problems. Section 4C deals with discrete data problems, and in particular those involving the binomial and Poisson distributions. Finally, in Section 4D we present a variety of problems not falling into any of the earlier categories, nor into the specialised types of inference discussed in Chapters 3 and 5.

4A One Sample: Normal Distribution

In this first section we confine attention to the case in which a single random sample is selected from a normal distribution, and inferences are required about any unknown parameters. Using the conventional notation $N(\mu, \sigma^2)$, the possible problems are those of making inferences about μ when σ is known (of very limited applicability in practice, but useful as a stepping-stone on the way to dealing with more realistic cases) and of making inferences about one or both of the parameters when neither is known. (The case where μ is known but σ is not rarely occurs in practice.) In each of these cases an inference may be required either in the form of a test of an appropriate hypothesis or in the form of the production of a confidence interval.

Notation for inferences relating to normal distributions is virtually standardised, and we shall be using it freely here and elsewhere. In particular, $\bar{x}$ (or, alternatively, $\bar{y}$, $\bar{x}_2$, etc.) will always be a sample mean, and s^2 denotes a sample variance (using the divisor $n - 1$).

Notation for percentage points of the normal, t, χ^2 and F-distributions is not as standardised, and it is convenient to note our conventions here. For the standardised normal distribution, the symbol z seems to be emerging as a near-standard, and a simple and convenient notation is $z_{0\cdot05}$ for the 5% point, $z_{0\cdot01}$ for the 1% point, and so on; thus $z_{0\cdot05}$ is the point *above* which 5% of $N(0,1)$ lies, i.e. $z_{0\cdot05} = 1\cdot645$. When working algebraically, we denote the point above which a proportion α lies (sometimes known as the 100α percentage point, or written as the $100\alpha\%$ point) by z_α.

A similar notation will be used for the other distributions, except that t and χ^2 have a parameter, the number of 'degrees of freedom', attached to them; the F-distribution has two such parameters. We use subscripts to denote the numbers of degrees of freedom, so that, for example, the 5% point of the t-distribution on 8 degrees of freedom will be written $t_{8,0\cdot05}$; its value is $1\cdot860$. Further examples follow; the 1% point of the χ^2 distribution on 10 degrees of freedom (which takes the value 23·21) will be written as $\chi^2_{10,0\cdot01}$, and the 0·1% point of the F-distribution on 3 and 7 degrees of freedom (18·77) will be denoted by $F_{3,7,0\cdot001}$. (When we work algebraically, we modify the notation in a natural way; note that when the number of degrees of freedom is a function it will sometimes be placed in brackets, for clarity, as in $t_{(n-1),\alpha/2}$, the point in the t-distribution on $n-1$ degrees of freedom above which a proportion $\alpha/2$ lies.)

4A.1 Lengths of manufactured bolts

Bolts are manufactured with a nominal length of 5 cm, and it is known from past experience that the variance of the lengths of such bolts is $0\cdot05\ \text{cm}^2$. A random sample of 10 bolts is taken from a box containing a large number of bolts, and their lengths (in cm) are found to be

5·68, 5·13, 5·82, 5·71, 5·36, 5·52, 5·29, 5·77, 5·45, 5·39.

(a) Find 95% confidence limits for the mean length, μ, of bolts in the box, stating clearly any assumptions made in deriving the limits.

(b) Without doing a formal test of a hypothesis, discuss whether $\mu = 5$ cm is a plausible hypothesis, given the result in part (a).

(c) By looking only at the number of bolts, out of 10, whose lengths are greater than 5 cm, construct a formal test of the null hypothesis $H_0: \mu = 5$ cm, against the alternative $H_1: \mu \neq 5$ cm.

Solution

(a) We make the following assumptions.

(i) The variance, σ^2, of the lengths of the bolts is known, and is $0\cdot05\ \text{cm}^2$.

(ii) The observations are normally distributed, so that the sample mean $\bar{X}$ has the distribution $N(\mu, \sigma^2/n)$, i.e. $N(\mu, 0\cdot005)$ when $\sigma^2 = 0\cdot05$ and $n = 10$.

With these assumptions, 95% confidence limits are given by $\bar{x} \pm 1\cdot96\sigma/\sqrt{n}$. For the present data, $\bar{x} = 5\cdot512$, so the limits are

$$5\cdot512 \pm 1\cdot96\sqrt{0\cdot05/10}$$

or $5\cdot512 \pm 0\cdot139$, i.e. the limits are 5·373 and 5·651. (Another way of expressing this is to say that (5·373, 5·651) is a 95% confidence interval for μ.)

(b) The value $\mu = 5$ cm is a long way outside the 95% confidence limits for μ, and therefore $\mu = 5$ cm does not seem to be a plausible value for the parameter μ.

(c) Each bolt is independently chosen, and each has the same probability of having a length greater than 5 cm. Thus the number Y of bolts longer than 5 cm is a binomial random variable with the number of 'trials' equal to 10. The probability p of success depends upon μ; if $\mu = 5$ cm and the distribution is symmetric (though not necessarily normal) then $p = \frac{1}{2}$. Thus, making the assumption of symmetry, the null hypothesis $H_0: \mu = 5$ cm is equivalent to $H_0: p = \frac{1}{2}$, and a two-sided alternative will be equivalent to $H_1: p \neq \frac{1}{2}$. The test statistic is simply Y, and the probability of obtaining a value of Y as extreme as the observed value, $y = 10$, is

$$\Pr(Y = 10) + \Pr(Y = 0) = (\tfrac{1}{2})^{10} + (\tfrac{1}{2})^{10} = (\tfrac{1}{2})^9 = 0{\cdot}002,$$

because a two-tailed test is required. Since the probability found is less than 0·01, H_0 would be rejected at the 1% level (also, of course, at the 5% and 10% levels) but not at the 0·1% level.

Notes

(1) The statement of the problem clearly implies that the variance is to be treated as known (i.e. that assumption (i) should be made). However, suppose now that σ had not been given, and so had to be treated as unknown. Then the expression $\bar{x} \pm z_{\alpha/2}\sigma/\sqrt{n}$ for the confidence limits would need to be replaced by $\bar{x} \pm t_{(n-1),\alpha/2}\, s/\sqrt{n}$, where s^2 is the sample variance and $t_{(n-1),\alpha/2}$ is the value exceeded with probability $\alpha/2$ by a random variable with the t-distribution on $n-1$ degrees of freedom, i.e. it is the appropriate percentage point of t.

In the present example,

$$s^2 = \frac{1}{n-1}\left\{\sum_{i=1}^{n} x_i^2 - \left(\sum_{i=1}^{n} x_i\right)^2 \Big/ n\right\} = \frac{1}{9}\left\{304{\cdot}2874 - (55{\cdot}12)^2/10\right\} = 0{\cdot}051\,77,$$

and $t_{9,0{\cdot}025} = 2{\cdot}26$, so the limits become $5{\cdot}512 \pm 2{\cdot}26\times\sqrt{0{\cdot}051\,77/10}$, or $5{\cdot}512 \pm 0{\cdot}163$; i.e. the limits of the 95% confidence interval are 5·349 and 5·675.

The value of s^2 is very similar to the value specified earlier for σ^2, but the limits for μ are wider, since t-distributions have longer tails than does $N(0,1)$. This reflects the additional uncertainty because of the fact that σ is unknown, rather than being a fixed, known, constant.

Assumption (ii), that observations are normally distributed, is needed because the sample size used is rather small. When the sample size is large, the sample mean $\bar{X}$ is approximately normally distributed, regardless (almost!) of the distribution of the individual observations, from the Central Limit Theorem. However, the result is asymptotic, and cannot be assumed to be effective when the sample size is as small as 10, unless the shape of the distribution of individual observations is known to be very close to that of the normal distribution.

(2) There is an equivalence between tests of hypotheses and confidence intervals. In the present example, a test of $H_0: \mu = \mu_0$ against $H_1: \mu \neq \mu_0$ will have as a test statistic

$$Z = \frac{\bar{X} - \mu_0}{\sigma/\sqrt{n}},$$

and H_0 will be rejected in a test of size (or significance level) α if the value z taken by Z is such that $|z| \geq z_{\alpha/2}$. But z will be in this rejection region if, and only if, μ_0 is outside the corresponding confidence interval for μ with confidence coefficient $1-\alpha$.

Since $\mu_0 = 5$ cm is outside the 95% confidence interval found in part (a) it follows that $H_0: \mu = 5$ cm will be rejected at the 5% level. In fact,

$$z = \frac{5{\cdot}512 - 5{\cdot}000}{\sqrt{0{\cdot}05/10}} = 7{\cdot}24,$$

so that H_0 will in this case be rejected also at very much smaller significance levels.

(3) The solution given to part (c) is not the only acceptable one, since in principle other types of test could be constructed. However, little could be done without making an assumption of symmetry. The test used in the solution is, in fact, known as the *sign test*, and is a non-parametric test. Since the test converts the hypothesis into one concerning p, the probability that an observation is greater than 5 cm, the null hypothesis is then expressible as

$$\Pr(\text{observation} > 5) = \tfrac{1}{2},$$

and one is really testing the hypothesis that the population *median* is 5 cm. When the distribution is symmetric, the mean and median are, of course, identical.

4A.2 Lifetimes of electrical components

Consider a confidence interval, with confidence coefficient $1-\alpha$, for the mean of a normal distribution with known variance σ^2, based on a random sample of n observations. How does the width of the interval change

(i) as n is increased, keeping σ^2 and α fixed;

(ii) as σ^2 is increased, keeping n and α fixed;

(iii) as α is decreased, keeping n and σ^2 fixed?

The standard deviation of the lifetime of a certain type of electrical component is 144 hours. How large a sample of the components must be taken to be (a) 95%, (b) 99% confident that the error in the estimated mean lifetime of such components will not exceed (i) 15 hours, (ii) 20 hours?

Solution

The width of the confidence interval is $2z_{\alpha/2}\sigma/\sqrt{n}$, so that

(i) as n increases, the width decreases and is proportional to $1/\sqrt{n}$;

(ii) as σ^2 increases, the width increases and is proportional to σ;

(iii) as α decreases, the width increases, since the width is proportional to $z_{\alpha/2}$ and $z_{\alpha/2}$ increases as α decreases, i.e. as $1-\alpha$, the confidence coefficient, increases.

These three results are all intuitively reasonable.

We now assume that lifetimes are normally distributed, and that the term 'error in the estimated mean' is interpreted as the half-width, w, of an appropriate confidence interval. (Other interpretations are discussed in Note 4.) Because of normality, $w = z_{\alpha/2}\sigma/\sqrt{n}$, since the confidence interval has end-points $\bar{x} \pm z_{\alpha/2}\sigma/\sqrt{n}$, and we require $w \leq w_0$, where $w_0 = 15, 20$ in (i), (ii) respectively.

Now $w \leq w_0$ implies $z_{\alpha/2}\sigma/\sqrt{n} \leq w_0$ or $n \geq (z_{\alpha/2}\sigma/w_0)^2$. We are given that $\sigma = 144$ and from tables of the normal distribution we find $z_{\alpha/2} = 1{\cdot}96$ and $2{\cdot}58$ in (a) and (b) respectively. Thus we have, rounding up to the nearest integer (see Note 6),

$$\text{for case (a)(i),}\quad n \geq \left[\frac{1{\cdot}96\times 144}{15}\right]^2 \approx 355,$$

$$\text{for case (a)(ii),}\quad n \geq \left[\frac{1{\cdot}96\times 144}{20}\right]^2 \approx 200,$$

$$\text{for case (b)(i),}\quad n \geq \left[\frac{2{\cdot}58\times 144}{15}\right]^2 \approx 614,$$

$$\text{for case (b)(ii),}\quad n \geq \left[\frac{2{\cdot}58\times 144}{20}\right]^2 \approx 346.$$

As might be expected, larger sample sizes are needed to achieve higher degrees of confidence, and also to obtain narrower confidence intervals.

Notes

(1) The assumption of normality is almost certainly intended here, even though distributions of lifetimes are often positively skewed and hence not normal. In any case, with the large values of n found in the solution, the mean of the observations will be approximately normally distributed (by the Central Limit Theorem) even if individual observations are not. There are also possibilities for transforming data before an analysis to reduce skewness. See, for example, the data in Problems 3B.5 and 4B.5, where logarithms of the original values were used.

(2) In the solution we assumed normality of the observations. No alternative distribution seems very natural, although general results can be obtained from Tchebychev's inequality, which is valid for any distribution. The inequality states that, for any random variable Y,

$$\Pr(|Y - \mathrm{E}(Y)| > k\sqrt{\mathrm{Var}(Y)}\,) < \frac{1}{k^2},$$

where k is any positive constant. If $Y = \bar{X}$, then its variance is σ^2/n, and $k\sigma/\sqrt{n}$ can be interpreted as the 'error in the estimated mean', since it is the half-width of a confidence interval derived from Tchebychev's inequality.

For a 95% confidence interval, put $\dfrac{1}{k^2} = 0{\cdot}05$, so that $k = 4{\cdot}47$. Then $k\sigma/\sqrt{n} \le w_0$ implies that

$$n \ge k^2\sigma^2/w_0^2 = 20\sigma^2/w_0^2.$$

For case (a)(i), $w_0 = 15$ and $\sigma = 144$, so that $n \ge 20\times(144)^2/15^2 = 1843$. Similar calculations may be done for the other parts of the problem.

Because Tchebychev's inequality holds for *any* distribution, the confidence interval which it gives for μ is generally much wider than a corresponding confidence interval based on specific distributional assumptions. Hence the sample size needed to achieve a given width has to be much greater when it is based on Tchebychev's inequality, rather than on specific distributional assumptions.

(3) Another interpretation of the problem is that we are required to find the minimum value of n such that $\Pr(|\bar{X} - \mu| \le w_0) \ge 1 - \alpha$. However, since $\bar{X} \sim N(\mu, \sigma^2/n)$, it follows that $Z = \dfrac{\bar{X} - \mu}{\sigma/\sqrt{n}} \sim N(0,1)$, and $\Pr(|\bar{X} - \mu| \le w_0) = \Pr\left(|Z| \le \dfrac{w_0\sqrt{n}}{\sigma}\right)$. For this probability to be greater or equal to $1 - \alpha$, we must have $\dfrac{w_0\sqrt{n}}{\sigma} \ge z_{\alpha/2}$, i.e. $n \ge \left(\dfrac{z_{\alpha/2}\sigma}{w_0}\right)^2$, so we have the same answer as before.

(4) Other interpretations of 'error in the estimated mean' include the full width of a confidence interval for the mean, and the standard error of the mean. The full width of the interval is $2w$, so that the required sample sizes will be 2^2, or 4, times those for the half interval, w. By contrast, the standard error of the mean is $\sigma/\sqrt{n} = w/z_{\alpha/2}$, so for that interpretation the required sample sizes will be $(1/z_{\alpha/2})^2$ times those for w.

(5) If the variance, σ^2, is not known, but an estimate s^2 is available based on an initial (pilot) sample of size m, then $n \ge \left\{t_{(m-1),\alpha/2}\dfrac{s}{w_0}\right\}^2$ will provide a guide to the size of sample needed in order to provide a confidence interval of half-width less than or equal to w_0. However, this will only be an approximate guide to the correct value for n. Usually the estimate of σ^2 used in the confidence interval will be based on all the n observations taken, and will be different from s^2, unless $n \le m$, in which case no further observations are needed.

If $n > m$, the estimated value for n will be approximate, but it is more likely to overestimate than to underestimate since $t_{(m-1),\alpha/2} > t_{(n-1),\alpha/2}$ for $n > m$.

(6) For case (a)(i), $(z_{\alpha/2}\sigma/w_0)^2 = 354{\cdot}04$. If $n = 354$, then w is (very slightly) greater than w_0. But n must be an integer, since it is a sample size, so we need $n \geq 355$ in order to obtain $w \leq w_0$. Similar considerations apply, of course, in the three other cases.

4A.3 The diameters of marbles

A machine produces marbles whose diameters are normally distributed with mean 12·00 mm. After modification of the machine, the diameters of a random sample of 105 marbles produced were found to have mean 12·010 mm and standard deviation 0·050 mm. Would you conclude that the modification has affected the mean diameter?

Solution

Since the population standard deviation is unknown, a z-test cannot be used, and a t-test is appropriate. The null hypothesis is that the modification has not affected the mean diameter, i.e. that $\mu = 12{\cdot}00$ mm; the alternative is that the mean has been affected, i.e. that $\mu \neq 12{\cdot}00$ mm, a two-sided alternative. We note that $\bar{x} = 12{\cdot}010$, that $s = 0{\cdot}050$ and that $n = 105$.

The t-statistic is defined as

$$t = \frac{\bar{x} - \mu}{s/\sqrt{n}}, \qquad (*)$$

so here

$$t = \frac{12{\cdot}010 - 12{\cdot}000}{0{\cdot}05/\sqrt{105}} = 2{\cdot}049,$$

and this must be compared with percentage points of the t-distribution on $n-1$, i.e. 104, degrees of freedom. Since the alternative hypothesis is two-sided, a two-tailed test is needed, and the two-tailed 5% point of t_{104} is 1·98, so the result is just significant at the 5% level. Such a level is commonly regarded as providing moderate, if not strong, evidence against a null hypothesis; we conclude that the mean diameter has been affected.

Notes

(1) Had the problem referred to the modification increasing rather than affecting the mean diameter, the alternative hypothesis would have been one-sided, i.e. that $\mu > 12{\cdot}00$ mm, and a one-tailed test would have been required. The t-statistic is still 2·049, but the 5% critical value is now 1·66. The hypothesis is still rejected at the 5% level but not at the 1% level, for which the critical value is 2·36.

(2) With 104 degrees of freedom, the t-statistic can be considered as approximately distributed as $N(0,1)$. So in this case, as in others when the sample size exceeds about 30, the more easily memorable percentage points of $N(0,1)$, e.g. 1·96 for a two-tailed 5% point, may be used as an approximation. The conclusion is the same.

(3) The problem might easily have asked for a 95% confidence interval to be given for μ, instead of for a test of a hypothesis. The confidence interval argument would have started from the fact that, using equation (*), the statistic t has Student's t-distribution on 104 degrees of freedom, and that only μ (the parameter of interest) is unknown. The two-tailed 5% point of t_{104} being 1·98, we therefore have

$$\Pr\left(-1{\cdot}98 \leq \frac{\bar{x} - \mu}{s/\sqrt{n}} \leq 1{\cdot}98\right) = 0{\cdot}95,$$

or

$$\Pr(\bar{x} - 1{\cdot}98s/\sqrt{n} \leq \mu \leq \bar{x} + 1{\cdot}98s/\sqrt{n}) = 0{\cdot}95.$$

The 95% confidence interval for μ is thus $(\bar{x} - 1{\cdot}98s/\sqrt{n}\,,\ \bar{x} + 1{\cdot}98s/\sqrt{n})$ and substituting $\bar{x} = 12{\cdot}010$, $s = 0{\cdot}050$, $n = 105$ gives the answer, viz. (12·000, 12·020). The null hypothesis value 12·00 seems to come just on the edge of this confidence interval, which reflects the fact that the t-statistic in the solution was very close to the critical value 1·98. However, if we were to work to unrealistic accuracy, we would find the lower limit of the confidence interval to be just above 12·00. The exclusion (if only just!) of this point $\mu = 12{\cdot}00$ from the 95% two-sided interval is merely a re-expression of the statement that on a 5% two-tailed test the hypothesis $\mu = 12{\cdot}00$ is rejected. (See also Note 2 to Problem 4A.1.)

(4) Although it is correct to conclude, as in the solution, that the mean diameter has been affected, most practising statisticians would take note of the fact that the value 2·049 is only just larger than 1·98, the critical value. Had one of the sample values been only a little smaller, the result might not have been significant at the 5% level, and statisticians are, generally, averse to basing decisions on evidence as slender as this. In such cases one might recommend that further evidence be obtained with a view to clarifying the matter. (Note that if a second random sample of 105 also had mean 12·010 mm and s.d. 0·050 mm, the value of t from the combined sample of 210 would now be 2·90, and would be judged highly significant.)

(5) The use by statisticians of the word 'significant' gives rise to much confusion. The difficulty lies in the fact that the word has slightly different technical and non-technical meanings, and not unnaturally these can be confused.

To focus ideas we consider the example of a crop experiment in which the aim is to compare the effects of two fertilisers A and B, applying each fertiliser to a random sample of plants and using a t-test to compare the sample mean yields. From the 'significance test' one might decide that 'the mean yield for A is significantly higher than that for B'.

Now the natural interpretation of this is that A is better than B by an amount which is 'significant', or important, so that A would be likely to give much better results than B. Unfortunately, such an interpretation would be wrong. The true meaning of 'significant', in this context, is simply that the observed result *signifies* that the null hypothesis is felt to be implausible and is thus rejected. In other words, the difference between the sample means is sufficiently large to indicate to us that the true mean for A is greater than that for B; note, though, that we have not said by how much it is greater, and hence we cannot say that A is better by an important, or 'significant', amount. Indeed, one reason for using large samples is to enable us to detect very *small* differences.

As we have seen, a hypothesis test cannot tell us by how much the mean for A exceeds that for B. This question will often be of interest, and the complementary technique of confidence intervals aims to answer it.

4A.4 Variability in weight of bottled fruit

The manager of a bottling plant is anxious to reduce the variability in net weight of fruit bottled. Over a long period the standard deviation has been 15·2 gm. A new machine is introduced, and the net weights (in grams) in randomly selected bottles (all of the same nominal weight) are

987, 966, 955, 977, 981, 967, 975, 980, 953, 972.

Would you report to the manager that the new machine has a better performance?

Solution

With data of this sort we assume that the weights are independently normally distributed, say $N(\mu, \sigma^2)$. The mean μ is unknown, and interest centres on the variance σ^2, for which we have $H_0 : \sigma^2 = \sigma_0^2 = 15{\cdot}2^2$; $H_1 : \sigma^2 < \sigma_0^2$. For tests concerning the variance of a normal distribution the appropriate statistic is $C = (n-1)s^2/\sigma_0^2$, where s^2 is the sample variance; denoting the 10 sample members by $x_1, \ldots, x_{10}$, we have $\Sigma x_i = 9713$ and $\Sigma x_i^2 = 9\,435\,347$, and hence $\Sigma(x_i - \bar{x})^2 = 1110{\cdot}1$. This is, of course, $(n-1)s^2$, so $C = 4{\cdot}805$. Under the null hypothesis

C has the χ^2 distribution with 9 degrees of freedom, and the lower 5% point is 3·30. The null hypothesis that the standard deviation is unchanged at 15·2 gm cannot therefore be rejected, so we would not report that the new machine is better.

Notes

(1) The χ^2 distribution is asymmetric, and many people find using its percentage points tricky. In the problem above it is clear that the lower – and less usual – tail is needed, since we wish to accept the alternative hypothesis that the standard deviation has been reduced only if the sample standard deviation is sufficiently small. In working out what values of χ^2 constitute the lower tail, it may be helpful to remember that $E(C) = \nu$, where ν is the number of degrees of freedom. So in the case being discussed, values of χ^2 above 9 are towards the upper tail, values less than 9 towards the lower tail. (A fuller discussion of this point is given in Note 4 to Problem 3B.2.)

(2) Suppose the problem had asked for a two-sided 95% confidence interval for σ^2. To obtain a confidence interval from first principles, one starts with a *pivotal function*, a function of sample members and the parameter of interest alone, whose distribution is known. In the present case, $C = (n-1)s^2/\sigma^2$ is such a function, and its distribution is indeed known, i.e. χ^2 with 9 degrees of freedom. (Another example is given in Problem 4B.6.) Using tables of the χ^2 distribution, we can thus write

$$\Pr\left\{2{\cdot}70 \leq \frac{(n-1)s^2}{\sigma^2} \leq 19{\cdot}02\right\} = 0{\cdot}95,$$

with equal tail areas being omitted at each end of the range. The confidence interval for σ^2 is thus

$$\left(\frac{(n-1)s^2}{19{\cdot}02}, \frac{(n-1)s^2}{2{\cdot}70}\right).$$

Since $(n-1)s^2 = 1110{\cdot}1$, the interval is (58·36, 411·15). The corresponding 95% confidence interval for σ, the standard deviation, is found by simply taking square roots, leading to the interval (7·64, 20·28).

(3) With many modern calculators it is quite feasible to work directly with raw data (as we did above) when obtaining sums and sums of squares, but one must naturally be careful to avoid errors resulting from truncation, particularly when finding the square of a number with many significant digits. For the data above, we might well have coded by subtracting 950, say, and indeed this would have been very desirable had the data been analysed without using a calculator.

4A.5 Changes in weight of manufactured product

(a) For a certain type of manufactured product, the weight is expected to be 350 gm, and the standard deviation is known to be 20 gm. The mean weight of 9 randomly selected items was 362 gm. Was there evidence of a change in weight, and if so at what level of significance?

(b) The process was changed, but with the standard deviation remaining the same. Nine randomly selected items had weights (in grams) of

327, 350, 374, 359, 397, 367, 331, 368, 385.

Is there now evidence that the mean weight has been increased?

(c) For the data in part (b), would it be reasonable to suggest that the standard deviation has probably changed?

Solution

(a) The population standard deviation (σ, say) is known to be 20, so that a t-test is unnecessary. We test the null hypothesis that the mean μ is $\mu_0 = 350$ against a two-sided alternative. The sample mean $\bar{x}$ is 362, and we calculate

$$z = \frac{\bar{x} - \mu_0}{\sigma/\sqrt{n}} = \frac{362 - 350}{20/\sqrt{9}} = 1{\cdot}80.$$

This lies between the familiar percentage points 1·645 and 1·96, so the result is significant at the 10% level only. We conclude that the evidence for a change is only very slight.

(b) The natural estimate of μ is provided again by the sample mean $\bar{x}$, which is again 362. The test statistic z is, thus, still 1·80, but the wording of the problem shows that this time a one-tailed test is needed. The 5% point is therefore 1·645, and we would conclude that there is reasonable evidence for an increase in mean weight.

(c) We now need to test the hypothesis $\sigma = 20$ against a two-sided alternative hypothesis. The test of a general null hypothesis $\sigma = \sigma_0$ is based on the statistic $C = (n-1)s^2/\sigma_0^2$, which under this null hypothesis has a χ^2 distribution on $n - 1$ degrees of freedom, n being the sample size. Here $n = 9$ and (coding the observations by subtracting 300 from each) the sum of the observations is 558 and the sum of squares is 38 894. Hence

$$(n-1)s^2 = 38\,894 - 558^2/9 = 4298.$$

The test statistic C is thus $4298/20^2$, or 10·745, to be compared with tables of χ^2 on 8 degrees of freedom. Now $\chi^2_{8,0{\cdot}9} = 3{\cdot}49$ and $\chi^2_{8,0{\cdot}1} = 13{\cdot}4$, so the value observed lies comfortably in the central portion of the distribution. We do not, therefore, reject the null hypothesis $\sigma = 20$, and we conclude that the evidence does not suggest that the standard deviation has probably changed.

Notes

(1) All the analyses undertaken in this problem assume that the observations are a random sample from a normal distribution. Such assumptions are commonly left unstated, as in the formal solution here, but it is really a good idea always to make assumptions explicit. Of the assumptions made here, it would not be possible to assess randomness without information as to how sampling was conducted. However, assuming randomness, normality can in principle be tested informally using a probability plot, or more formally using a goodness-of-fit test such as the χ^2 test. The problems in Section 3B show how such tests can be done, though with a sample of size only 9 it would be almost impossible to determine whether the underlying distribution is normal.

(2) Notwithstanding the comments in Note 1, z-tests and t-tests are not very sensitive to failure of an assumption of normality, in the sense that if the distribution is actually not normal but of a not utterly dissimilar shape the size (i.e. the significance level) of a test will be close to the nominal size. This result follows from the Central Limit Theorem. By contrast, tests on variances are distinctly more sensitive to non-normality, and F-tests (for comparing two sample variances) are extremely sensitive.

(3) This problem is based on a real examination question. But the analysis required in part (b), in particular, falls short of normal statistical practice. It is hard to imagine circumstances in which, a change having occurred, one could be so sure that the standard deviation was unaltered that one would wish to analyse data without first checking the assumption. In practice one would usually avoid such an assumption by using a t-test.

4A.6 Estimating the mean and variance

(a) Discuss the two expressions $\frac{1}{n}\sum_{i=1}^{n}(x_i-\bar{x})^2$ and $\frac{1}{(n-1)}\sum_{i=1}^{n}(x_i-\bar{x})^2$, both of which are used to measure the spread of a set of observations $x_1, x_2, \ldots, x_n$.

(b) A random sample of n observations is taken from a distribution; the sum of the observations is t_1, and the sum of the squares of the observations is t_2. Explain how to estimate the mean and the variance of the distribution from which the random sample was taken.

(c) Given the random sample described in part (b), write down expressions (based on t_1 and t_2) for estimates of the mean and variance of the mean of a further, independent, random sample of size m, from the original distribution.

(d) Given that $n = 25$, $t_1 = 400$ and $t_2 = 8800$, construct a 99% confidence interval for the mean of the distribution, and use it to test whether or not this mean could be 20.

Solution

(a) Given a set of n observations, $x_1, x_2, \ldots, x_n$, the quantity $s_1^2 = \frac{1}{n}\sum_{i=1}^{n}(x_i-\bar{x})^2$ gives a measure of 'spread' for this set of observations. It can be thought of as the variance of the discrete probability distribution which assigns probability $1/n$ to each of the values $x_1, x_2, \ldots, x_n$. (Further discussion can be found in Note 8 to Problem 3A.1.) If we simply wish to summarise the 'spread' of observations, then s_1^2, or its square root, provides a suitable measure.

If the n observations are a random sample from some larger (possibly infinite, possibly hypothetical) population or distribution, then we may want to use the sample to make inferences about the variance, σ^2, of the population or distribution. The statistic s_1^2 can be used as an estimate of σ^2, as well as being a summary measure for the sample, but $s^2 = \frac{1}{(n-1)}\sum_{i=1}^{n}(x_i-\bar{x})^2$ is usually preferred as an estimate. There are various reasons for this preference (and some reasons, too, why s_1^2 might be used) but the main one is that s^2 is obtained from an unbiased estimator for σ^2, whereas s_1^2 is not. Of course, if n is large, it makes very little practical difference whether s^2 or s_1^2 is used.

(b) The obvious estimate of the mean, μ, is t_1/n. Also, following on from part (a), the obvious estimate of the variance σ^2 is s^2, which can be written as

$$\begin{aligned} s^2 &= \frac{1}{(n-1)}\sum_{i=1}^{n}(x_i-\bar{x})^2 \\ &= \frac{1}{(n-1)}\left\{\sum_{i=1}^{n}x_i^2 - 2\bar{x}\sum_{i=1}^{n}x_i + n\bar{x}^2\right\} \\ &= \frac{1}{(n-1)}\left\{\sum_{i=1}^{n}x_i^2 - \left(\sum_{i=1}^{n}x_i\right)^2/n\right\} \\ &= \frac{1}{(n-1)}\left\{t_2 - \frac{t_1^2}{n}\right\}. \end{aligned}$$

(c) If the original distribution has mean μ and variance σ^2, then the sample mean, for a sample of size m, has a distribution with mean μ and variance σ^2/m. The obvious estimates of these quantities are, from (b), $\bar{x} = \frac{t_1}{n}$ and $\frac{s^2}{m} = \frac{1}{m(n-1)}\left\{t_2 - \frac{t_1^2}{n}\right\}$.

(d) Assuming that the distribution is normal, confidence limits for the mean are of the form $\bar{x} \pm t_{(n-1),\alpha/2}\frac{s}{\sqrt{n}}$. Now $\bar{x} = \frac{400}{25} = 16{\cdot}0$ and $s^2 = \frac{1}{24}\left[8800 - \frac{(400)^2}{25}\right] = 100$. Also $t_{24,0{\cdot}005} = 2{\cdot}80$, from tables, so the limits of the 99% confidence interval are $16{\cdot}0 \pm 2{\cdot}80 \times 10/\sqrt{25}$, or $16{\cdot}0 \pm 5{\cdot}6$. Thus the required interval is (10·4, 21·6).

This interval includes the value 20·0, so on the basis of this confidence interval the hypothesis that $\mu = 20$ would be accepted at the 1% level.

Notes

(1) No assumptions are asked for, or spelled out, in this problem. Either the assumption of normality is intended, or perhaps it could be assumed that, in part (d), $n = 25$ is large enough for the sample mean to be normally distributed, regardless of the distribution of individual observations. In either case the expression $\bar{x} \pm z_{\alpha/2}s/\sqrt{n}$ might be used for the confidence interval, that is, we might replace $t_{(n-1),\alpha/2}$ by $z_{\alpha/2}$. For $n = 25$ and $\alpha = 0{\cdot}01$, we have $t_{24,0{\cdot}005} = 2{\cdot}80$. But $z_{0{\cdot}005} = 2{\cdot}58$, so the corresponding interval will be somewhat narrower.

(2) The value 20 hypothesised for μ is not very far inside the interval. For example, if a 90% interval is constructed, rather than a 99% interval, using $t_{24,0{\cdot}05} = 1{\cdot}71$, then the interval is (12·58 , 19·42), which does not include 20. Thus the hypothesis $\mu = 20$ would have been rejected at the 10% level. (See Note 2 to Problem 4A.1, amongst other places, for further discussion of the links between confidence intervals and tests of hypotheses.)

4B Two Samples: Normal Distribution

We now move on to the naturally more complex case where we have observations from two normal distributions, and the object is to compare corresponding parameters of the distributions. In many respects the analyses of means are, mathematically, straightforward extensions of single-sample procedures; but there are techniques needed for two-sample problems which do not derive directly from the single-sample case. For example, when comparing variances of two samples the appropriate analysis requires a test based on the F-distribution.

4B.1 Weights of tins of peas

(a) On a particular day a random sample of 12 tins of peas is taken from the output of a canning factory, and their contents are weighed. The mean and standard deviation of weight for the sample are 301·8 gm and 1·8 gm respectively. Find 99% confidence limits for the mean weight of peas in tins produced by the factory on the day in question.

(b) On the following day a further random sample of 12 tins is taken, and the mean and standard deviation of contents for this sample are 302·1 gm and 1·6 gm respectively. Assuming that the variances of the weights are the same on the two days, show that a 95% confidence interval for the difference between mean weights on the two days includes zero.

(c) Assume now that the samples on both days are from the same population. Find a 99% confidence interval for the mean weight of tins in that population, based on both samples.

Solution

(a) The confidence limits for the mean net weight of tins produced on the first day have the general form

$$\bar{x} \pm t_{(n-1),\alpha/2}\frac{s}{\sqrt{n}},$$

as was given in part (d) of the solution to Problem 4A.6. In the current example, $\bar{x} = 301{\cdot}8$, $s = 1{\cdot}8$, $n = 12$, $\alpha = 0{\cdot}01$, and $t_{11,0{\cdot}005} = 3{\cdot}106$, so the limits are

$$301{\cdot}8 \pm 3{\cdot}106 \times \frac{1{\cdot}8}{\sqrt{12}} = 301{\cdot}8 \pm 1{\cdot}61.$$

The limits are therefore 300·19 gm and 303·41 gm.

(b) The phrasing of this part of the question differs from that of part (a), in that it asks for a confidence *interval* rather than confidence *limits*. Confidence limits are simply the end-points of a confidence interval. Making the suggested assumption of equal variances for the populations of tins produced on the two days, the expression for a confidence interval for the difference in means between the two days has the general form

$$(\bar{x}_1 - \bar{x}_2) \pm t_{(n_1+n_2-2),\alpha/2}\, s_p \left\{\frac{1}{n_1} + \frac{1}{n_2}\right\}^{\frac{1}{2}},$$

where s_p^2 is the pooled variance for the two samples (see Note 3), and the remaining notation is conventional.

But $\bar{x}_1 = 301{\cdot}8$, $\bar{x}_2 = 302{\cdot}1$, $s_p^2 = 2{\cdot}90$, $\alpha = 0{\cdot}05$, and $t_{22,0{\cdot}025} = 2{\cdot}074$, so the required interval has end-points

$$-0{\cdot}3 \pm 2{\cdot}074 \times \sqrt{2{\cdot}90} \times \sqrt{\frac{1}{12} + \frac{1}{12}}$$

$$= -0{\cdot}3 \pm 2{\cdot}074 \times \sqrt{2{\cdot}90/6} = -0{\cdot}3 \pm 1{\cdot}44.$$

The interval is therefore $(-1{\cdot}74, 1{\cdot}14)$, which includes zero.

(c) Since the confidence interval for the difference between mean weights includes zero the assumption specified, that the samples come from a common population, seems not unreasonable. Given two independent random samples of sizes n_1 and n_2, with sample means $\bar{x}_1$ and $\bar{x}_2$ and with sample variances s_1^2 and s_2^2, the mean and variance of the combined sample are given by

$$\bar{x} = \frac{n_1\bar{x}_1 + n_2\bar{x}_2}{n_1 + n_2}$$

and

$$s^2 = \frac{(n_1 - 1)s_1^2 + (n_2 - 1)s_2^2 + \dfrac{n_1 n_2(\bar{x}_1 - \bar{x}_2)^2}{(n_1 + n_2)}}{n_1 + n_2 - 1}.$$

(See Problem 4B.3 for a fuller discussion of these formulae.) For the present data,

$$\bar{x} = \frac{(12 \times 301{\cdot}8) + (12 \times 302{\cdot}1)}{24} = 301{\cdot}95,$$

and

$$s^2 = \frac{(11 \times (1{\cdot}8)^2) + (11 \times (1{\cdot}6)^2) + \dfrac{144 \times (0{\cdot}3)^2}{24}}{23} = 2{\cdot}80.$$

The required confidence interval has the same form as that in part (a), but with $\bar{x}$, s and n now taking the values just calculated for the combined sample of 24 observations. Thus the interval has end-points $\bar{x} \pm t_{23,0{\cdot}005} \times s/\sqrt{n}$, which become $301{\cdot}95 \pm 2{\cdot}81 \times 1{\cdot}67/\sqrt{24}$ or $301{\cdot}95 \pm 0{\cdot}96$. The interval is therefore (300·99, 302·91). (An alternative, simpler, interval is discussed in Note 3.)

Notes

(1) In each part of this problem, a confidence interval or pair of confidence limits is required. Occasionally a single confidence limit, or equivalently a one-sided confidence interval, is wanted and the current problem illustrates a situation where a one-sided interval might be considered appropriate.

Suppose that the tins are labelled as weighing 300 gm net, and that it is a legal requirement for the mean weight of contents of all tins to be at least 300 gm. The management will then be more interested in a lower limit for the mean weight than an upper limit (though, of course, if the mean weight is too far above 300 gm, profits will be eroded). A lower limit, with confidence coefficient $1-\alpha$, is

$$\bar{x} - t_{(n-1),\alpha}\, s/\sqrt{n}.$$

In the current example, $t_{11,0\cdot01} = 2\cdot718$, so that a 99% lower confidence limit for the mean weight, based on the first sample, is

$$301\cdot8 - (2\cdot718\times1\cdot8/\sqrt{12}),$$

or 300·39 gm. The corresponding one-sided confidence interval now consists of all values greater than 300·39 gm. Although this type of interval is substantially different from the two-sided 99% interval, in particular because it has no upper limit, it still has the same probability of covering the true mean weight. In fact, any interval of the form

$$\left(\bar{x} - t_{(n-1),\alpha_1}\frac{s}{\sqrt{n}},\ \bar{x} + t_{(n-1),\alpha_2}\frac{s}{\sqrt{n}}\right),$$

where $\alpha_1 + \alpha_2 = \alpha$, has the same coverage probability $(1-\alpha)$. In practice, however, the most usual choices of α_1, α_2 are

(i) $\alpha_1 = \alpha_2 = \alpha/2$, leading to the usual (two-sided) confidence interval,

(ii) $\alpha_1 = \alpha$, $\alpha_2 = 0$, leading to a lower confidence limit, and a corresponding one-sided confidence interval with no upper bound, and

(iii) $\alpha_1 = 0$, $\alpha_2 = \alpha$, leading to an upper confidence limit, and a corresponding one-sided confidence interval with no lower bound.

The equivalence between (two-sided) confidence intervals and (two-tailed) tests of hypotheses was discussed in Note 2 to Problem 4A.1. There is similarly an equivalence between one-sided confidence intervals and one-tailed tests of hypotheses. For example, a test of H_0: $\mu = \mu_0$ against H_1: $\mu > \mu_0$ (or H_1: $\mu < \mu_0$) will reject H_0 at significance level α if and only if μ_0 is below (above) the one-sided lower (upper) confidence limit for μ, with confidence coefficient $1-\alpha$.

Finally, it should be noted that one-sided confidence limits, and intervals, can readily be found in other situations (e.g. intervals for variances, binomial parameters, differences between means, etc.) by simple modifications of the expressions for two-sided intervals. Often, all that is needed is to look at one, rather than both, of the two-sided limits, but replace $\frac{1}{2}\alpha$ by α in the cut-off of the appropriate distribution (z, t, χ^2, etc.).

(2) For the confidence interval in part (b) to be valid, one must assume that variances for the two populations of interest are equal. There is also an unspoken assumption that weights are normally distributed underlying all the confidence intervals in the problem.

If the assumption of equal variances is in doubt, it can be tested using the F-test (see the Note to Problem 4B.3). In the present case s_1^2 and s_2^2 are close enough for a formal test to be unnecessary but, for completeness, we perform the test. The ratio of sample variances is easily seen to be

$$\frac{s_1^2}{s_2^2} = \frac{(1\cdot8)^2}{(1\cdot6)^2} = \frac{3\cdot24}{2\cdot56} = 1\cdot27.$$

This is to be compared with percentage points of the F-distribution on n_1-1 and n_2-1 degrees of freedom and, since $F_{11,11,0\cdot05} = 2\cdot82 > 1\cdot27$, we would accept the null hypothesis that the two variances are equal. (The result is not even significant at the 10% level, on a two-tailed test.)

When the variances are not equal, there is no single standard form for a confidence interval for the difference in means, unless n_1 and n_2 are large. In this event an interval whose confidence coefficient is approximately $1-\alpha$ is given by

$$(\bar{x}_1 - \bar{x}_2) \pm z_{\alpha/2}\left\{\frac{s_1^2}{n_1} + \frac{s_2^2}{n_2}\right\}^{\frac{1}{2}},$$

using a natural notation. Since the two samples are large, the sample variances will be good estimates of the true variances, so the expression is an approximation to

$$(\bar{x}_1 - \bar{x}_2) \pm z_{\alpha/2}\left\{\frac{\sigma_1^2}{n_1} + \frac{\sigma_2^2}{n_2}\right\}^{\frac{1}{2}},$$

which gives a confidence interval, with confidence coefficient exactly $1-\alpha$, when the variances σ_1^2 and σ_2^2 are known, and normality can be assumed. The latter expression is still approximately valid in the absence of normality, for large enough samples, because of the Central Limit Theorem.

Returning to the expression

$$(\bar{x}_1 - \bar{x}_2) \pm z_{\alpha/2}\left\{\frac{s_1^2}{n_1} + \frac{s_2^2}{n_2}\right\}^{\frac{1}{2}}$$

we illustrate its use for the current data. The sample sizes are really too small to guarantee a good approximation but, because s_1^2 and s_2^2 are very similar, the interval given by the approximation will not be too different, numerically, from that found earlier. The interval has end-points

$$-0\cdot3 \pm 1\cdot96\left\{\frac{(1\cdot8)^2}{12} + \frac{(1\cdot6)^2}{12}\right\}^{\frac{1}{2}}$$

$$= -0\cdot3 \pm 1\cdot96\times0\cdot695,$$

i.e. the interval is $(-1\cdot66, 1\cdot06)$. In fact, because of the equal sample sizes,

$$\left\{\frac{s_1^2}{n_1} + \frac{s_2^2}{n_2}\right\} = s_p^2\left\{\frac{1}{n_1} + \frac{1}{n_2}\right\},$$

where s_p^2 is the pooled variance, as used in part (b), and the only difference between the interval found there and the present one is the replacement of $t_{22,0\cdot025} = 2\cdot074$ by $z_{0\cdot025} = 1\cdot96$, which makes the latter interval somewhat narrower than it should be.

(3) The so-called pooled variance, s_p^2, for the two samples is given by

$$s_p^2 = \frac{(n_1-1)s_1^2 + (n_2-1)s_2^2}{n_1+n_2-2},$$

which reduces to $\frac{1}{2}(s_1^2 + s_2^2)$ when $n_1 = n_2$. Note that

$$s_p^2 = \frac{\sum_{i=1}^{n_1}(x_{1i} - \bar{x}_1)^2 + \sum_{i=1}^{n_2}(x_{2i} - \bar{x}_2)^2}{n_1+n_2-2}$$

which is, in general, not the same as $s^2 = \sum_{j=1}^{2}\sum_{i=1}^{n_j}\frac{(x_{ji} - \bar{x})^2}{n_1+n_2-1}$, the sample variance which would be obtained if we treated all n_1+n_2 observations as coming from a single population, as required in the final part of the problem. In part (b), the population variances, but clearly not the means, are assumed to be the same, so we use the first expression.

In part (c), a confidence interval can be constructed using s_p^2 instead of s^2. Such an interval has end-points

$$\bar{x} \pm t_{(n_1+n_2-2),\alpha/2}\frac{s_p}{\sqrt{n}}.$$

This will generally be slightly wider than that based on s, which uses the additional information that the two samples are drawn from populations with the same mean, and hence gains a degree of freedom. However, the interval is only slightly wider, and is somewhat easier to calculate. In the present example, the interval using s_p^2 is (300·97, 302·93), with width 1·96, compared to the interval using s^2 which has width 1·92.

4B.2 Experimental teaching of arithmetic

In a study of a new method of teaching arithmetic, claimed to improve on conventional techniques, 400 children were divided at random into two groups A and B, of sizes 250 and 150 respectively. Those in group B were taught using the experimental method, while those in group A were taught by conventional methods. After completion of the course all the pupils were given the same test paper, and the group scores were as follows.

Group A : mean 67·8, variance 60·4,
Group B : mean 70·2, variance 55·6.

Is there any evidence that method B is effective? State carefully your null hypothesis, alternative hypothesis and conclusion.

Solution

We use a natural notation, with a subscript A or B indicating the group concerned. So we have $n_A = 250$, $n_B = 150$, $\bar{x}_A = 67{\cdot}8$, $\bar{x}_B = 70{\cdot}2$, $s_A^2 = 60{\cdot}4$, and $s_B^2 = 55{\cdot}6$. The $n_A = 250$ observations on method A will be assumed to be independent $N(\mu_A, \sigma_A^2)$, and similarly the $n_B = 150$ on B are $N(\mu_B, \sigma_B^2)$. The null hypothesis will be that the two methods are equally effective; i.e. we have $H_0: \mu_A = \mu_B$. Since B is an experimental method a natural alternative to consider is whether it improves on the standard method, A, so we choose the one-sided alternative hypothesis $H_1: \mu_A < \mu_B$.

Since the population variances σ_A^2 and σ_B^2 are unknown, a natural test to consider is a two-sample t-test, which depends for its validity on equality of these two variances. In this case s_A^2 and s_B^2 are very close, so such an assumption is very reasonable. The appropriate test statistic for the hypothesis H_0 is then

$$t = \frac{\bar{x}_A - \bar{x}_B}{s\left(\dfrac{1}{n_A} + \dfrac{1}{n_B}\right)^{\frac{1}{2}}},$$

where s^2, the pooled estimate of the common population variance, is given by

$$s^2 = \frac{(n_A - 1)s_A^2 + (n_B - 1)s_B^2}{(n_A - 1) + (n_B - 1)},$$

and the statistic t has the t-distribution on $n_A + n_B - 2$ degrees of freedom.

Straightforward calculation now gives

$$s^2 = \frac{249\times60{\cdot}4 + 149\times55{\cdot}6}{398} = 58{\cdot}60,$$

so that

$$t = \frac{67{\cdot}8 - 70{\cdot}2}{\sqrt{58{\cdot}60}\left(\dfrac{1}{250} + \dfrac{1}{150}\right)^{\frac{1}{2}}} = -3{\cdot}036.$$

The t-distribution on 398 degrees of freedom is, for all practical purposes, the same as $N(0,1)$, so for a one-tailed test at the 5% level the critical value is $-1{\cdot}645$. Corresponding critical values for 1% and 0·1% tests are $-2{\cdot}326$ and $-3{\cdot}090$. We therefore reject the hypothesis at the 1% level (if not quite at the 0·1% level), and conclude that there is strong evidence that method B improves on its competitor.

Notes

(1) The plausibility of the assumption that $\sigma_A^2 = \sigma_B^2$ can be judged formally by setting up that assumption as a hypothesis and testing it. The test for equality of variances of two normal distributions is an F-test with test statistic s_A^2/s_B^2, here equal to 1·086. This has to be compared with percentage points of the F-distribution on 249 and 149 degrees of freedom, using a two-tailed test (unless there is some reason to do otherwise, not hinted at in the statement of the present problem).

Unfortunately, few published tables deal in such high numbers for the degrees of freedom, but one can sometimes get a lower bound for a percentage point by looking at the '∞' row or column of tables of F. So, for example, we find that the 10% point (on a two-tailed test) of $F_{\infty,150}$ is 1·223, and the corresponding point of $F_{249,149}$ will be larger than this. Since the observed ratio is only 1·086, the hypothesis of equality of σ_A^2 and σ_B^2 cannot be rejected. (It is possible to be more precise than this, since there are formulae, quoted in some books of tables, giving approximate percentage points for use when the numbers of degrees of freedom are large.)

(2) In the solution we assumed the data to be normally distributed when deriving the t-test. In practice, when samples are as large as they are here, the Central Limit Theorem gives assurance that the sample means have distributions adequately approximated by normal distributions, unless the distribution of the individual observations is very odd.

(3) While arguments based on the t-test are logically sound, there is an alternative two-sample test for equality of normal means. This is available when the samples are large (i.e. the theory justifying it is asymptotic) but it does have the advantage of not requiring the assumption that $\sigma_A^2 = \sigma_B^2$. The equivalent confidence interval procedure is described in Note 2 to Problem 4B.1.

(4) In the special case in which $n_A = n_B = n$, say, the test of the null hypothesis $\mu_A = \mu_B$ is carried out by comparing the statistic

$$\frac{\bar{x}_A - \bar{x}_B}{s\sqrt{2/n}}$$

with percentage points of the t-distribution on $2n-2$ degrees of freedom. Now if a two-tailed test is required, one can perform the test by comparing $|\bar{x}_A - \bar{x}_B|$ with $ts\sqrt{2/n}$, where t represents the appropriate percentage point of the t-distribution. What we are doing is, therefore, to compare the observed difference between the sample means with

$$t_{\nu,\alpha/2}\sqrt{2s^2/n},$$

where ν is the number of degrees of freedom, here $2n-2$. The quantity displayed is known as the Least Significant Difference, and an equivalent quantity is used in Problems 5B.1 and 5B.3.

4B.3 Milk yield of cows

(a) Two independent random samples of sizes n_1, n_2 are available, with sample means $\bar{x}_1$, $\bar{x}_2$ and sample variances s_1^2, s_2^2 respectively. If the two samples are combined to give a single sample of size $n_1 + n_2$, and with sample mean $\bar{x}$ and variance s^2, show that

$$\bar{x} = \frac{n_1\bar{x}_1 + n_2\bar{x}_2}{n_1 + n_2}$$

and

$$s^2 = \frac{(n_1 - 1)s_1^2 + (n_2 - 1)s_2^2 + \dfrac{n_1 n_2}{n_1 + n_2}(\bar{x}_1 - \bar{x}_2)^2}{n_1 + n_2 - 1}.$$

(b) A random sample of 10 dairy cows is taken from a large herd at Farm A, and the weekly milk yield, in kg, is recorded for each cow in the sample. Similar measurements are made for a random sample of 15 cows from a large herd at Farm B. The sample mean and sample variance for Farm A are 142 kg and 440 kg^2, whereas the sample mean and sample variance for the combined sample of 25 cows are 158 kg and 816 kg^2 respectively. Calculate the mean and variance for the sample of cows from Farm B.

(c) Find a 95% confidence interval for the variance of milk yield at Farm A.

(d) Repeat part (c) for Farm B and discuss carefully, without doing any further formal analysis, whether the variances at the two farms could be equal.

Solution

(a) For the sample mean, we have

$$\begin{aligned}\bar{x} &= (\text{Sum of all } n_1 + n_2 \text{ observations})/(n_1 + n_2) \\ &= \left(\sum_{i=1}^{n_1} x_{1i} + \sum_{i=1}^{n_2} x_{2i}\right)/(n_1 + n_2), \text{ with obvious notation,} \\ &= (n_1\bar{x}_1 + n_2\bar{x}_2)/(n_1 + n_2).\end{aligned}$$

To derive the formula for sample variance we make use of an identity for the sum of squares of any set of observations $y_1, y_2, \ldots, y_n$ about any constant c, viz.

$$\sum_{i=1}^{n}(y_i - c)^2 = \sum_{i=1}^{n}(y_i - \bar{y})^2 + n(\bar{y} - c)^2, \qquad (*)$$

where $\bar{y}$ is the sample mean of the n observations.

From first principles, the sample variance of the $n_1 + n_2$ observations is given by

$$\begin{aligned}s^2 &= \frac{1}{n_1 + n_2 - 1}\{\text{Sum of squares about the common mean } \bar{x}\} \\ &= \frac{1}{n_1 + n_2 - 1}\left\{\sum_1^{n_1}(x_{1i} - \bar{x})^2 + \sum_1^{n_2}(x_{2i} - \bar{x})^2\right\}.\end{aligned}$$

Now, from equation (*), we have for the first sample

$$\sum_{i=1}^{n_1}(x_{1i} - \bar{x})^2 = \sum_{i=1}^{n_1}(x_{1i} - \bar{x}_1)^2 + n_1(\bar{x}_1 - \bar{x})^2 = (n_1 - 1)s_1^2 + n_1(\bar{x}_1 - \bar{x})^2,$$

and similarly for the second sample. Hence

$$(n_1 + n_2 - 1)s^2 = (n_1 - 1)s_1^2 + (n_2 - 1)s_2^2 + n_1(\bar{x}_1 - \bar{x})^2 + n_2(\bar{x}_2 - \bar{x})^2.$$

Now, since $\bar{x} = \dfrac{n_1\bar{x}_1 + n_2\bar{x}_2}{n_1 + n_2}$, we obtain

$$\bar{x}_1 - \bar{x} = \frac{n_2(\bar{x}_1 - \bar{x}_2)}{n_1 + n_2},$$

and similarly

$$\bar{x}_2 - \bar{x} = \frac{n_1(\bar{x}_2 - \bar{x}_1)}{n_1 + n_2}.$$

So $n_1(\bar{x}_1-\bar{x})^2 + n_2(\bar{x}_2-\bar{x})^2 = (\bar{x}_1 - \bar{x}_2)^2\left\{\dfrac{n_1n_2^2}{(n_1+n_2)^2} + \dfrac{n_1^2n_2}{(n_1+n_2)^2}\right\}$

$$= \frac{n_1n_2}{n_1+n_2}(\bar{x}_1 - \bar{x}_2)^2 .$$

Combining all these we reach the required result

$$s^2 = \frac{1}{n_1+n_2-1}\left\{(n_1-1)s_1^2 + (n_2-1)s_2^2 + \frac{n_1n_2}{n_1+n_2}(\bar{x}_1 - \bar{x}_2)^2\right\}.$$

(b) This is simply an application of part (a). For the means, we have (in kg)

$$\bar{x}_2 = \frac{(n_1 + n_2)\bar{x} - n_1\bar{x}_1}{n_2}$$

$$= \frac{25\times158 - 10\times142}{15} = 168{\cdot}67\,\text{kg}.$$

For variances,

$$s_2^2 = \frac{(n_1+n_2-1)s^2 - (n_1-1)s_1^2 - \dfrac{n_1n_2}{n_1+n_2}(\bar{x}_1-\bar{x}_2)^2}{n_2 - 1}$$

$$= \frac{24\times816 - 9\times440 - \dfrac{10\times15}{25}(142 - 168{\cdot}67)^2}{14}$$

$$= \frac{19584 - 3960 - 4266{\cdot}67}{14} = 811{\cdot}24\,\text{kg}^2.$$

(c) Confidence intervals for a variance are based on the χ^2 distribution; for example, the interval for σ_1^2 has the form

$$\frac{(n_1-1)s_1^2}{\chi^2_{(n_1-1),\alpha/2}} \le \sigma_1^2 \le \frac{(n_1-1)s_1^2}{\chi^2_{(n_1-1),(1-\alpha/2)}}$$

where $\chi^2_{(n_1-1),\alpha/2}$ and $\chi^2_{(n_1-1),(1-\alpha/2)}$ are upper and lower $\frac{1}{2}\alpha$ points, respectively, of the χ^2 distribution with (n_1-1) degrees of freedom. For a 95% interval, $\chi^2_{9,0{\cdot}025} = 19{\cdot}02$ and $\chi^2_{9,0{\cdot}975} = 2{\cdot}70$, from standard statistical tables, so the interval is

$$\left(\frac{9\times440}{19{\cdot}02}, \frac{9\times440}{2{\cdot}70}\right) = (208{\cdot}20,\ 1466{\cdot}67).$$

(d) The required interval is

$$\left(\frac{(n_2-1)s_2^2}{\chi^2_{(n_2-1),0\cdot025}}, \frac{(n_2-1)s_2^2}{\chi^2_{(n_2-1),0\cdot975}}\right),$$

and from a table of the χ^2 distribution we find that $\chi^2_{14,0\cdot025} = 26\cdot12$ and $\chi^2_{14,0\cdot975} = 5\cdot63$, so the interval is

$$\left(\frac{14\times811\cdot24}{26\cdot12}, \frac{14\times811\cdot24}{5\cdot63}\right) = (434\cdot81, 2017\cdot29)\,.$$

There is a considerable amount of overlap between the intervals for σ_1^2 and σ_2^2, so it seems plausible that the variances for the two farms could be the same.

* **Note**

Despite the conclusion above, it is not necessarily the case that overlap between confidence intervals implies that equality of the two parameters is plausible. Consider, for example, the normal means μ_1 and μ_2 when the corresponding variances σ_1^2, σ_2^2 are known. A confidence interval for $\mu_1 - \mu_2$ has half-width $z_{\alpha/2}(\sigma_1^2/n_1 + \sigma_2^2/n_2)^{\frac{1}{2}}$, (cf. Note 2 to Problem 4B.1) so equality of μ_1 and μ_2 is implausible if

$$|\bar{x}_1 - \bar{x}_2| > z_{\alpha/2}(\sigma_1^2/n_1 + \sigma_2^2/n_2)^{\frac{1}{2}}.$$

However, individual intervals for μ_1 , μ_2 will overlap if

$$|\bar{x}_1 - \bar{x}_2| < z_{\alpha/2}\sigma_1/\sqrt{n_1} + z_{\alpha/2}\sigma_2/\sqrt{n_2}.$$

Now $z_{\alpha/2}(\sigma_1/\sqrt{n_1} + \sigma_2/\sqrt{n_2}) \geq z_{\alpha/2}(\sigma_1^2/n_1 + \sigma_2^2/n_2)^{\frac{1}{2}}$. (In the simplest case when $\sigma_1 = \sigma_2$ and $n_1 = n_2$, the ratio of $(\sigma_1/\sqrt{n_1} + \sigma_2/\sqrt{n_2})$ and $(\sigma_1^2/n_1 + \sigma_2^2/n_2)^{\frac{1}{2}}$ is $\sqrt{2}$.) We see, therefore, that overlap between intervals can occur when $\mu_1 = \mu_2$ is implausible.

The formal test of H_0: $\sigma_1^2 = \sigma_2^2$ was not required here, but it is based on the F-distribution. H_0 is rejected in favour of H_1: $\sigma_1^2 \neq \sigma_2^2$ if

$$\frac{s_1^2}{s_2^2} > F_{(n_1-1),(n_2-1),\alpha/2} \quad \text{or} \quad \frac{s_2^2}{s_1^2} > F_{(n_2-1),(n_1-1),\alpha/2}.$$

(Since these percentage points of F are always greater than 1, we need only consider whichever of s_1^2/s_2^2 and s_2^2/s_1^2 is greater than one, and compare it with the relevant F-distribution.)

In the present example, $s_2^2/s_1^2 = 1\cdot84$ and $F_{14,9,0\cdot025} = 3\cdot80$, so H_0 would be accepted. This test can, in fact, be adapted to give a confidence interval for σ_1^2/σ_2^2 of the form

$$\frac{s_1^2/s_2^2}{F_{(n_1-1),(n_2-1),\alpha/2}} \leq \frac{\sigma_1^2}{\sigma_2^2} \leq \frac{s_1^2}{s_2^2}F_{(n_2-1),(n_1-1),\alpha/2}\,.$$

But $F_{9,14,0\cdot025} = 3\cdot21$, so the interval is

$$\left(\frac{1}{1\cdot84\times3\cdot21}, \frac{3\cdot80}{1\cdot84}\right) = (0\cdot17, 2\cdot07),$$

which comfortably includes the null value $\sigma_1^2/\sigma_2^2 = 1$.

4B.4 Speeding up mathematical tasks

Eight schoolchildren, chosen at random from the first year of a large school, were given, without prior warning, a mathematical task, and the time taken (in minutes) by each child to complete the task was recorded.

The following day the children were instructed how to perform such tasks efficiently, and a week later they were tested again on a similar task. Once again, the time taken to complete the task was recorded for each child and the results were as follows.

	Time taken (minutes)							
Child	1	2	3	4	5	6	7	8
Before instruction	26	20	17	21	23	24	21	18
After instruction	19	14	13	16	19	18	16	17

(a) Find a 90% confidence interval for the mean time taken by first year children (i) before instruction, and (ii) after instruction, assuming that times are normally distributed.

(b) Find a 90% confidence interval for the mean difference between times before and after instruction, for first year children.

(c) Approximately how many children would have been needed in the sample in order to achieve a confidence interval in part (b) whose total width is 2 minutes?

Solution

(a) The familiar formula $\bar{x} \pm t_{(n-1),\alpha/2}\frac{s}{\sqrt{n}}$ gives the required confidence intervals. We shall use a subscript 1 to denote measurements made before instruction and subscript 2 for the later ones. Before instruction,

$$\sum_{i=1}^{8} x_{1i} = 170, \ \sum_{i=1}^{8} x_{1i}^2 = 3676,$$

$$\text{so } \bar{x}_1 = \frac{170}{8} = 21{\cdot}25, \text{ and } s_1^2 = \frac{1}{7}\left\{3676 - \frac{(170)^2}{8}\right\} = \frac{63{\cdot}5}{7} = 9{\cdot}07.$$

(Coding the data would simplify the arithmetic a little – see Problem 3A.1.) After instruction,

$$\sum_{i=1}^{8} x_{2i} = 132, \ \sum_{i=1}^{8} x_{2i}^2 = 2212,$$

$$\text{so } \bar{x}_2 = \frac{132}{8} = 16{\cdot}5, \text{ and } s_2^2 = \frac{1}{7}\left\{2212 - \frac{(132)^2}{8}\right\} = \frac{34}{7} = 4{\cdot}86.$$

Since $t_{7,0\cdot05} = 1{\cdot}895$, the 90% confidence limits for the mean time before instruction are

$$21{\cdot}25 \pm 1{\cdot}895\sqrt{9{\cdot}07/8} = 21{\cdot}25 \pm 2{\cdot}02,$$

so the 90% confidence interval is (19·23 , 23·27). After instruction the corresponding limits become

$$16{\cdot}5 \pm 1{\cdot}895\sqrt{4{\cdot}86/8} = 16{\cdot}5 \pm 1{\cdot}48,$$

i.e. the 90% confidence interval is (15·02 , 17·98).

(b) To obtain an interval for the difference between means on the two occasions, we use the fact that the data are paired. The interval is then

$$\bar{d} \pm t_{n-1,\alpha/2}\frac{s_d}{\sqrt{n}},$$

where $\bar{d}$ is the mean difference for the samples, and s_d^2 is the sample variance for the individual differences. The individual differences are

$$7, 6, 4, 5, 4, 6, 5, 1,$$

so $\sum_{i=1}^{8} d_i = 38$ and $\sum_{i=1}^{8} d_i^2 = 204$. Hence $\bar{d} = \dfrac{38}{8} = 4{\cdot}75$ $(= 21{\cdot}25 - 16{\cdot}5 = \bar{x}_1 - \bar{x}_2)$, and

$$s_d^2 = \frac{1}{7}\left\{204 - \frac{(38)^2}{8}\right\} = \frac{23{\cdot}5}{7} = 3{\cdot}357.$$

The 90% confidence limits for the mean difference (or, equivalently, the difference between means) are therefore

$$4{\cdot}75 \pm 1{\cdot}895\sqrt{3{\cdot}357/8}$$

$$= 4{\cdot}75 \pm 1{\cdot}23.$$

The 90% confidence interval is thus (3·52, 5·98).

(c) The width of the confidence interval in part (b) is $2t_{(n-1),0\cdot05}\dfrac{s_d}{\sqrt{n}}$, and so it is proportional to $1/\sqrt{n}$. If we took a larger sample, then $t_{(n-1),0\cdot05}$ would decrease, and s_d would change, but to get an approximate idea of the sample size needed to achieve a given width, we will keep these quantities fixed.

A sample of size 8 gives an interval of width 2·46, so to get the width down to 2 (minutes), we will need a sample size of approximately $8\times(2{\cdot}46)^2/(2)^2 = 12{\cdot}1$, and rounding up gives a required sample size of 13 children.

Notes

(1) The expression for a confidence interval for a difference between two means used in Problem 4B.1 would not be appropriate here, because it does not take into account the correlation between the times for each child (those who were fastest before instruction also tend to be fastest after instruction). Because of this correlation, the quantity $s_p^2(1/n_1 + 1/n_2)$ overestimates the variance of $\bar{x}_1 - \bar{x}_2$, hence leading to an interval that is wider than necessary, as we shall now verify numerically. The erroneous interval is

$$(\bar{x}_1 - \bar{x}_2) \pm t_{(n_1+n_2-2),0\cdot05}\, s_p \sqrt{1/n_1 + 1/n_2}.$$

Now $t_{14,0\cdot05} = 1{\cdot}761$ and $s_p^2 = \dfrac{(n_1-1)s_1^2 + (n_2-1)s_2^2}{n_1+n_2-2} = \dfrac{63{\cdot}5 + 34}{14} = 6{\cdot}96$, so the apparent 90% confidence interval has end-points

$$4{\cdot}75 \pm 1{\cdot}761\left\{6{\cdot}96\left(\frac{1}{8} + \frac{1}{8}\right)\right\}^{\frac{1}{2}}$$

$$= 4{\cdot}75 \pm 1{\cdot}761\sqrt{6{\cdot}96/4} = 4{\cdot}75 \pm 2{\cdot}32,$$

and the interval is thus (2·43, 7·07). We see that the failure to take into account the paired nature of the data leads to a 'confidence interval' whose width is almost twice as great as that of the correct one. (A related matter is discussed in Note 1 to Problem 4B.5.)

(2) As with Problem 4B.1, the situation in this problem is one where a one-sided confidence interval might be of interest. Suppose, for example, that the instruction period involves expensive equipment. A lower confidence limit for the mean improvement in speed provided by the instruction might then be required before deciding on the purchase of expensive equipment.

As before, there is an equivalence between one-sided (two-sided) confidence intervals for differences between means and corresponding one-tailed (two-tailed) tests of hypotheses regarding differences in means. When *differences* between means (or between binomial proportions – see Problems 4C.5 and 4C.6) are of interest, it is much more common to be asked to set up a test of the hypothesis of no difference between means (or proportions), rather than to construct a confidence interval. The latter procedure is nevertheless a useful one, and it can be used to test the hypothesis of equal means (proportions) simply by seeing whether or not zero is contained in the interval.

* (3) If we wished to get a slightly better estimate of the required sample size, we could take account of the variation in $t_{(n-1),0\cdot05}$ as n varies, although it will only decrease from 1·895 to 1·645 as n goes from 8 to infinity, a relatively small amount of change compared with the possible variations in s_d as the sample size is increased. However, there is little we can do to take account of change in s_d, so we consider it to be fixed. The calculations in part (c) suggest a sample size of 13, but since the value of t should be less than 1·895, it may be possible to achieve the required width with a smaller sample size. In fact, for $n = 12$, the width is

$$2t_{11,0\cdot05}\times s_d/\sqrt{12} = 2\times1\cdot796\sqrt{3\cdot357/12} = 1\cdot90.$$

For $n = 11$, the width is

$$2\times1\cdot812\times\sqrt{3\cdot357/11} = 2\cdot002,$$

so the required width of 2 minutes is achieved, by these (still approximate) calculations when $n = 12$, but it is not quite achieved for $n = 11$. (It will, of course, be achieved for $n = 11$ if s_d for the larger sample is slightly smaller than that of the original sample. Conversely, it may not be achieved for larger n if s_d increases.)

4B.5 Peeling potatoes faster

An experiment was conducted to compare the performance of two potato peelers, and in particular to discover whether the typical user might be able to peel potatoes faster with one rather than the other. Ten volunteers were used, and each was given both peelers for a period before the experiment in order to gain familiarity with them. In the experiment the volunteer subjects used both peelers on standardised amounts of potatoes, and then repeated the experiment with the peelers used in the opposite order, so as to eliminate any effect due to ordering. The table below gives, for each subject, the mean of the natural logarithms of time (in seconds) needed to complete the tasks.

Subject	Peeler A	Peeler B
1	2·33	2·34
2	2·76	2·79
3	1·91	1·91
4	2·62	2·60
5	2·01	2·03
6	1·77	1·80
7	1·81	1·81
8	1·99	2·00
9	1·97	1·98
10	2·26	2·30

Use Student's t-test, at the 5% level of significance, to test whether the peelers differ in their efficiency.

Solution

For data of this type a paired sample t-test is required. We therefore form the 10 differences $d_1, \ldots, d_{10}$ between the results for peelers A and B, viz.

$$-0{\cdot}01,\ -0{\cdot}03,\ 0{\cdot}00,\ 0{\cdot}02,\ -0{\cdot}02,\ -0{\cdot}03\ ,\ 0{\cdot}00,\ -0{\cdot}01,\ -0{\cdot}01,\ -0{\cdot}04.$$

Treating them as a random sample from $N(\mu, \sigma^2)$, we test the null hypothesis $\mu = 0$ (with σ^2 unknown). In a natural notation, we thus compare

$$t = \frac{\bar{d} - 0}{s/\sqrt{n}}$$

with the two-tailed 5% points of Student's t-distribution on 9 degrees of freedom, i.e. with $\pm 2{\cdot}262$. Since $\bar{d} = -0{\cdot}013$ and $s = 0{\cdot}0177$, the value of t is $-2{\cdot}327$. We thus reject the null hypothesis, and conclude that the peelers differ in their mean level of efficiency.

Notes

* (1) For a paired-sample t-test to be justified, the differences must be a random sample from a normal distribution. A simple model which leads to the required condition is that

result = effect of subject + effect of peeler + random error,

as long as the random errors are themselves normally distributed. This model is in fact just a restatement of the standard model leading to a two-way analysis of variance, discussed in Problem 5B.3. In fact the paired-sample t-test is just a special case of two-way analysis of variance, though it is customarily justified using a more intuitive approach.

(2) The problem asks for a test that the two peelers differ. This does indicate, of course, that a two-tailed test is required, but is, as a hypothesis, too vague to use as a null hypothesis. (There is no way in which such a hypothesis could be formally disproved.)

(3) The paired sample test is used because the 20 observations arose through both peelers being used by the same 10 volunteers. Had there been 20 volunteers, 10 randomly allocated to peeler A and the remaining 10 to peeler B, the two-sample t-test (i.e. the test for two independent samples, as discussed in Problem 4B.2) would have been needed.

(4) If one peeler were more efficient than the other, one would expect it to have the effect of reducing the time required by a fraction rather than by a constant. Similarly, a good subject might do the job in half the time, rather than in 10 seconds less. For these reasons, it seems implausible for a model such as that in Note 1 to apply to the times themselves, but if one takes logarithms the model might well be a reasonable one.

There are many other circumstances in which one feels instinctively that a multiplicative rather than an additive model is likely to be more appropriate; the analysis then almost invariably starts by taking logarithms. This is particularly true in the area of economics, in which many quantities (inflation, wage claims, discounts) are naturally discussed in terms of percentages or proportions.

(5) The data used in this problem are recorded to 3 significant digits. But the test used is based on the differences $d_1, d_2, \ldots, d_{10}$, and through cancellation these are only available to one significant digit. In practice most statisticians would argue that in these circumstances the t-statistic could not be calculated at all accurately, and that the results from a test ought therefore to be regarded with some caution. In the solution the value of t was found to be $-2{\cdot}327$, and the appropriate percentage point was $-2{\cdot}262$. These values are really too close for us to feel confident in rejecting the null hypothesis at the 5% level.

Even if we were to reject the hypothesis, there remains the fact that the observed differences are very small. The discussion in Note 5 to Problem 4A.3 of the distinction between statistical and practical significance is of considerable relevance here.

* 4B.6 Estimating a ratio

Independent normal random variables X and Y have means μ and $\rho\mu$ respectively, and both have variance 1. The ratio ρ of their means is of particular interest. Show that the distribution of $Y - \rho X$ depends upon ρ, but not upon μ.

A single observation is made on each of X and Y, resulting in values x and y respectively. Use your result to test the hypothesis $\rho = 1$ against an alternative $\rho \neq 1$ at the 5% level when

(i) $x = 0{\cdot}3$ and $y = 2{\cdot}7$,

(ii) $x = 0{\cdot}3$ and $y = 1{\cdot}5$.

Solution

We are given that $X \sim N(\mu, 1)$ and $Y \sim N(\rho\mu, 1)$ independently. We thus obtain

$$\mathrm{E}(Y - \rho X) = \mathrm{E}(Y) - \rho \mathrm{E}(X) = \rho\mu - \rho\mu = 0$$

and, by independence,

$$\mathrm{Var}(Y - \rho X) = \mathrm{Var}(Y) + \rho^2 \mathrm{Var}(X) = 1 + \rho^2.$$

Since linear combinations of normally distributed random variables are themselves normally distributed we obtain

$$Y - \rho X \sim N(0, 1 + \rho^2),$$

depending on ρ alone, as required.

To test a hypothesis we examine the sampling distribution of an appropriate statistic, when the hypothesis is taken as true. The statistic $Y - \rho X$ is appropriate, since the unwanted parameter μ is eliminated, and under the null hypothesis that $\rho = 1$

$$Y - \rho X = Y - X \sim N(0, 2),$$

so that

$$\frac{1}{\sqrt{2}}(Y - X) \sim N(0, 1).$$

For case (i) the observed value of $Y - X$ is $y - x = 2{\cdot}4$, and $2{\cdot}4/\sqrt{2} = 1{\cdot}697$, which we must compare with the well-known percentage points of $N(0, 1)$, using a two-tailed test. Since $-1{\cdot}96 < 1{\cdot}697 < +1{\cdot}96$, we do not reject the hypothesis that $\rho = 1$ on the 5% two-tailed test, despite the observed ratio $2{\cdot}7/0{\cdot}3$ of $9{\cdot}0$.

We would naturally expect from this that for case (ii) the hypothesis will again not be rejected, and so it turns out. Now $y - x$ is $1{\cdot}2$, with a corresponding standardised deviate of $1{\cdot}2/\sqrt{2}$, which is clearly not outside the range $(-1{\cdot}96, +1{\cdot}96)$.

Notes

(1) This relatively unfamiliar piece of theory gives valuable insights. Note that the problem involves two unknown parameters μ and ρ; interest lies in one, ρ, only, while μ is an example of what is actually termed a *nuisance* parameter. In order to make inferences about the parameter of interest, μ must be eliminated from the problem. This is done in general by finding a so-called *pivotal* function: that is, a function of the observations and the parameter of interest whose sampling distribution does not depend on any parameter. In the present case $Y - \rho X$ has this property. (See Note 2 to Problem 4A.4 for a related discussion.)

Another, much more familiar, problem can also be set in the same framework, the problem of inference about a normal mean when the variance σ^2 is unknown. The mean of the distribution is now the parameter of interest, while σ is the nuisance parameter. As is well known, the problem is solved by use of Student's t-test. Had σ been known, then the appropriate statistic would have been

$$\frac{\bar{X} - \mu}{\sigma/\sqrt{n}},$$

using an obvious notation. However, with σ unknown, this statistic cannot be calculated, and we use instead the pivotal function

$$\frac{\bar{X} - \mu}{s/\sqrt{n}},$$

whose distribution (Student's t on $n - 1$ degrees of freedom) does not depend on the unknown parameter σ, and so overcomes the difficulty, just as $Y - \rho X$ does in the present problem.

(2) The problem also gives an unusual but most valuable insight into the meaning of confidence intervals. A common but mistaken view of a confidence interval is that the entire natural range for a parameter ($-\infty$ to ∞ in the case of ρ here) gives a 100% confidence interval for that parameter, while a lesser degree of confidence will be attached to any narrower interval. This simple approach does in fact work for many straightforward problems (and hence perpetuates the mistaken view!) but the present problem has attracted much academic interest because the simple ideas break down.

The easiest way of obtaining a confidence interval for ρ is to make use of the link with tests of hypotheses; a 95% confidence interval for ρ consists of all values for ρ not rejected by a hypothesis test at the 5% level. In the present problem we were required to test the hypothesis $\rho = 1$, and we can generalise this to test the hypothesis $\rho = \rho_0$. The appropriate test statistic is now $Y - \rho_0 X$, and when $\rho = \rho_0$ the distribution of the statistic is $N(0, 1 + \rho_0^2)$. Denoting the observed value of $Y - \rho_0 X$ by $y - \rho_0 x$, we see that we must compare

$$\frac{y - \rho_0 x}{\sqrt{1 + \rho_0^2}}$$

with percentage points of $N(0, 1)$, for example, with $\pm 1{\cdot}96$ for a 5% test. Accordingly a value ρ_0 for ρ lies in the 95% confidence interval if

$$-1{\cdot}96 \leq \frac{y - \rho_0 x}{\sqrt{1 + \rho_0^2}} \leq +1{\cdot}96,$$

i.e. if $\dfrac{(y - \rho_0 x)^2}{1 + \rho_0^2} \leq 1{\cdot}96^2 = 3{\cdot}8416$. The 95% confidence interval for ρ thus includes all those values ρ_0 satisfying

$$(y - \rho_0 x)^2 \leq 3{\cdot}8416(1 + \rho_0^2),$$

$$\text{or} \quad (y^2 - 3{\cdot}8416) - 2xy\rho_0 + (x^2 - 3{\cdot}8416)\rho_0^2 \leq 0.$$

(For any other confidence coefficient, one merely changes the coefficient $3{\cdot}8416 = 1{\cdot}96^2$.) One now obtains the confidence interval simply by solving a quadratic equation, but it is important to note that the confidence interval is not necessarily the interval *between* the two roots; if $|x| < 1{\cdot}96$ the coefficient of ρ_0^2 is negative, so values between the roots will be the only ones *not* in the confidence interval! This happens, of course, when $x = 0{\cdot}3$, so in case (i) here the 95% confidence interval for ρ is the entire real line *except* for the interval $(-1{\cdot}20, 0{\cdot}77)$. Note that the value $\rho = 1$ lies inside the confidence interval, as is required, since the hypothesis $\rho = 1$ was not rejected.

One would have expected the same feature to occur in case (ii), but in fact this case is still more awkward. When $x = 0{\cdot}3$, $y = 1{\cdot}5$, the quadratic inequality reduces to

$$-1{\cdot}5916 - 0{\cdot}9\rho_0 - 3{\cdot}7516\rho_0^2 \leq 0,$$

$$\text{or} \quad 3{\cdot}7516\rho_0^2 + 0{\cdot}9\rho_0 + 1{\cdot}5916 \geq 0.$$

The corresponding quadratic equation has no real roots, so the inequality is satisfied for *all* real values ρ_0. It follows that the 95% confidence interval for ρ is $(-\infty, \infty)$, a finding which gives the amateur and the experienced statistician alike rather a jolt.

The paradox arises because it is easy to slip into the error of imagining that every 95% confidence interval has a 95% chance of covering the true value of the parameter concerned, and similarly for other levels of confidence, so that $(-\infty, \infty)$ can only be a 100% confidence interval. Of several correct interpretations of confidence intervals, it is convenient to use here the link with hypothesis tests. A 95% confidence interval is that interval including all values of the parameter held to be consistent with the data (i.e. not rejected) on a 5% test. In the present case (ii), it is clear that it would be hard to sustain a claim that any particular value for $\rho = E(Y)/E(X)$ was incompatible with the two observations. A simple (and, frankly, too rough-and-ready) argument to justify this looks at x and y separately. From $x = 0{\cdot}3$, a 95% confidence interval for $\mu = E(X)$ is $(-1{\cdot}66, 2{\cdot}26)$. A similar interval for $\rho\mu = E(Y)$ is $(-0{\cdot}46, 3{\cdot}46)$. Since both intervals include positive and negative values, and since in particular the interval for $E(X)$ contains 0 it is easy to see that it is quite plausible for ρ to be small or large, positive or negative; the information available from x and y does not enable us to exclude any real value from the confidence interval as being incompatible with the data.

We have perhaps taken the reader quite far enough into uncharted territory already. We ought, however, to add the comment that this paradoxical – not to say perverse – behaviour of confidence intervals noted above has not escaped the attention of academic statisticians, and that it has been the subject of lively controversy.

4C Binomial and Poisson Distributions

While calculations for tests and intervals are generally simplest for normal distributions, the problems are no less important when other distributions are involved, most particularly the binomial and Poisson. (In fact these distributions can be especially instructive, particularly when the numbers involved are kept small and convenient.)

A difficulty with the binomial and Poisson distributions is that, with increasing sample size, calculations quickly become tiresome, especially when finding confidence intervals. Approximations are then of great value, and in particular the normal distribution can provide a very good approximation, as mentioned in Note 1 to Problem 2A.3 and in Problem 2A.7. Use of the normal distribution is illustrated several times in this section.

4C.1 The pink and white flowers

A seed manufacturer claims that in a particular variety sold by him there will be one white flower for every three pink flowers. You buy a packet and plant the contents, obtaining 21 pink and 3 white flowers. Do you accept the manufacturer's claim?

Solution

The 'claim' that 75% of flowers will be pink is simply a hypothesis that p, the probability of a pink flower, is 0·75. The number X of pink flowers arising from the planting of 24 seeds is binomially distributed; so we are required to examine whether 21 seeds is 'extreme' in the context of $B(24, 0{\cdot}75)$. The significance level of the result is just the probability of observing a result at least as extreme. Now the probability of observing 21 or more is

$$\binom{24}{21}(0{\cdot}75)^{21}(0{\cdot}25)^3 + \binom{24}{22}(0{\cdot}75)^{22}(0{\cdot}25)^2 + \binom{24}{23}(0{\cdot}75)^{23}(0{\cdot}25)^1 + (0{\cdot}75)^{24},$$

or 0·1150. This would, then, be the significance level on a one-tailed test, but here a two-tailed test is needed, since there is nothing in the problem to suggest that an alternative $p > 0{\cdot}75$ is intended. The two-tailed test will have a significance level roughly double $0{\cdot}115$, so there is no question of rejecting the hypothesis $p = 0{\cdot}75$. We accordingly accept the manufacturer's claim.

Notes

(1) Two-tailed tests for discrete data are difficult, as they are with non-symmetrical sampling distributions generally, since it is hard to judge when a result is as extreme as the one observed, but in the other tail. In the current case a two-tailed test is needed, and the difficulty just mentioned is avoided because there is no question of significance even on a one-tailed test, and thus *a fortiori* the two-tailed test will be non-significant.

Had a significance level been required, the most reasonable solution would have been to treat the (binomial) distribution as symmetrical about $x = 18$. (Note 1 to Problem 2A.3 provides some justification for this.) We would then calculate

$$\Pr(X \leq 15) + \Pr(X \geq 21)$$

as the significance level. In fact $\Pr(X \leq 15) = 0{\cdot}1213$, so overall the significance level is $0{\cdot}1150 + 0{\cdot}1213 = 0{\cdot}2363$.

(2) For the $B(24, 0{\cdot}75)$ distribution a normal approximation is not unreasonable. Using a continuity correction, we have

$$\Pr(X \geq 21) \approx 1 - \Phi\left(\frac{21 - \frac{1}{2} - 18}{\sqrt{9/2}}\right) = 1 - \Phi(1{\cdot}1785) = 0{\cdot}1193.$$

Note that the approximate value $0{\cdot}1193$ lies between the two exact values for the two tails, $0{\cdot}1150$ and $0{\cdot}1213$.

(3) A 3 : 1 ratio of pink to white flowers arises from Mendelian theory when colour is determined by a single gene with two alleles, pink being dominant over white, and when cross-breeding occurs in a particular way. Probability theory is frequently used by geneticists. It is nowadays being used to advantage in genetic counselling, when parents with a family history of some hereditary disease contemplate having a child.

Conversely, genetic applications offer practical interest to the student of probability, since the probabilities involved can be calculated exactly, rather than just algebraically. To obtain numerical results gives some satisfaction, but these are generally only available when one can identify equally likely outcomes for the experiment involved. At an elementary level, at least, such outcomes can rarely be identified outside trivial games of chance, rather dry problems of sampling and problems in genetics.

4C.2 Distinguishing between brands of cheese

It is claimed that 90% of men cannot tell the difference between two different brands of Cheddar cheese, but of the members of a random sample of 500 men, 72 could distinguish between them. Is the claim justified?

Solution

Let p be the proportion of the male population who can distinguish, and let X be the number in the sample who can. Reasonable assumptions indicate that X has the binomial distribution with index 500, i.e. $n = 500$, and parameter p. The null hypothesis is that $p = 0{\cdot}1$, and under this hypothesis $\mathrm{E}(X) = 50$, whereas in fact 72 were observed.

The significance level of the test is $\Pr(X \geq 72)$, where this probability is calculated under the assumption that the null hypothesis $X \sim B(500, 0{\cdot}1)$ is true. For $n = 500$, $p = 0{\cdot}1$, the normal approximation to the binomial distribution is quite adequate, so we treat X as if $X \sim N(50, 45)$. Using the continuity correction, we thus calculate

$$z = \frac{72 - \frac{1}{2} - 50}{\sqrt{45}} = 3{\cdot}205,$$

and we require the significance level of such a value. Consulting a normal table, we find that the value is above the 0·1% point (3·09) on a one-tailed test, and we therefore confidently reject the claim.

Notes

(1) The 'reasonable' assumptions referred to in the solution are that the results from different sample members are independent, each having the same probability of being able to distinguish between the cheeses. These conditions are assured once we are told that a random sample has been selected, although if one were undertaking a real investigation there are more questions to be asked, for example about just what the target population is.

(2) In the solution, a one-tailed test was used, though the wording of the question does not unambiguously indicate the need for this. It seems reasonable to argue, however, that the claim is really that *at least* 90% cannot tell the difference, since no-one would contemplate rejecting the claim if it turned out that 95% cannot distinguish. Indeed, another way of expressing the hypotheses involved in this question is to say that one is testing $H_0: p \leq 0{\cdot}1$ against $H_A: p > 0{\cdot}1$.

(3) Use of the normal approximation to the binomial distribution is standard when the numbers involved are of the order of magnitude of those in this problem. (Notes 1 and 2 to Problem 2A.3 give further discussion of the approximation.)

Had the numbers involved been smaller, the logic of the test would have been the same, but use of the normal distribution would not have been justified. Had the sample been of size 10, with 3 being able to distinguish (clearly unrealistically small numbers but adequate for displaying principles) then $X \sim B(10, p)$ and the significance level would have been $\Pr(X \geq 3)$, when $p = 0{\cdot}1$. The probability is most easily calculated as

$$\begin{aligned} &1 - \Pr(X = 0) - \Pr(X = 1) - \Pr(X = 2) \\ &= 1 - (0{\cdot}9)^{10} - 10(0{\cdot}9)^9(0{\cdot}1) - 45(0{\cdot}9)^8(0{\cdot}1)^2 \\ &= 0{\cdot}070. \end{aligned}$$

Such a result would therefore have been significant at the 10% level but not at the 5% level, and would not have been regarded as providing strong evidence against the claim.

4C.3 Women investors in building societies

A building society wishes to estimate the proportion, p, of its savings account holders who are women. Records are scanned for a sample of 80 accounts and it is found that 22 are held by women.

(a) Find an approximate two-sided 90% confidence interval for p, stating clearly any assumptions which are necessary for the approximation to be valid.

(b) Ideally, the society would like to estimate p with a precision of $\pm 0{\cdot}02$ (in the sense that a 90% confidence interval should have half-width no greater than 0·02). How large a sample size would be necessary to achieve this precision?

Solution

(a) We use the normal approximation to the binomial distribution, i.e. if X is the number of women in a sample of size n, and $\hat{p} = X/n$, then $\hat{p} \sim N(p, p(1-p)/n)$, with the usual notation. The two assumptions which are necessary for this result are as follows.

(i) X has a binomial distribution, for which we need the sample to be random. This implies that all accounts have the same chance of being chosen, and also that the probability of an account-holder being female does not depend on the sex of the other account-holders in the sample. This can only be *approximately* true for a *finite* population of account-holders, but is a reasonable approximation provided that the sample is only a small fraction of the population.

(ii) The sample size is large enough for the normal approximation to be an adequate approximation to the binomial. A value of 80 is certainly large enough, unless p is close to zero or one, which is not the case here. (Further discussion of this point can be found in Note 1 to Problem 2A.3.)

Returning to the approximate result $\hat{p} \sim N(p, p(1-p)/n)$, it follows that

$$\Pr\left\{\left|\frac{\hat{p}-p}{\sqrt{p(1-p)/n}}\right| \le z_{\alpha/2}\right\} = 1-\alpha.$$

This leads to two possible ways of obtaining an approximate confidence interval for p (cf. Problem 4D.4 for the corresponding possibilities for a Poisson parameter). Replacing p by $\hat{p}$ in the denominator of the left hand side of the inequality leads to

$$\Pr\left\{\hat{p} - z_{\alpha/2}[\hat{p}(1-\hat{p})/n]^{\frac{1}{2}} \le p \le \hat{p} + z_{\alpha/2}[\hat{p}(1-\hat{p})/n]^{\frac{1}{2}}\right\} = 1-\alpha,$$

so that an approximate confidence interval for p, with confidence coefficient $1-\alpha$, has end-points

$$\hat{p} \pm z_{\alpha/2}[\hat{p}(1-\hat{p})/n]^{\frac{1}{2}}.$$

In the present example, $\hat{p} = \frac{22}{80} = 0{\cdot}275$, and, for a 90% interval, $z_{0\cdot05} = 1{\cdot}645$, so the limits are

$$0{\cdot}275 \pm 1{\cdot}645\{0{\cdot}275\times0{\cdot}725/80\}^{\frac{1}{2}} = 0{\cdot}275 \pm 0{\cdot}082.$$

The interval is therefore $(0{\cdot}193, 0{\cdot}357)$.

A second possibility is to square the inequality

$$\left|\frac{\hat{p}-p}{\sqrt{p(1-p)/n}}\right| \le z_{\alpha/2},$$

obtaining the quadratic inequality

$$p^2(n + z_{\alpha/2}^2) - p(2n\hat{p} + z_{\alpha/2}^2) + n\hat{p}^2 \le 0.$$

Since the coefficient of p^2 is positive, the inequality will be satisfied for values of p lying between the two roots of the corresponding quadratic equation, and these roots will therefore give the end-points of an approximate confidence interval for p. Given the large sample size, the results for the confidence interval are unlikely to be very different from those obtained by the cruder approximation given above, but we give them for the sake of completeness. For the present problem we have $n = 80$, $\hat{p} = \frac{22}{80}$ and $z_{\alpha/2} = 1{\cdot}645$, so the quadratic equation becomes

$$82{\cdot}706p^2 - 46{\cdot}706p + 6{\cdot}050 = 0,$$

and the roots of this equation are $0{\cdot}201$ and $0{\cdot}363$, i.e the interval is $(0{\cdot}201, 0{\cdot}363)$.

As predicted, the interval is similar to that given by the cruder approximation, but note that it is not centred on the sample proportion $\hat{p}$. The lower limit is closer to $\hat{p}$ than is the upper limit; this is standard behaviour for $\hat{p} < \frac{1}{2}$, and the phenomenon, unsurprisingly, becomes more extreme as $\hat{p}$ approaches zero. Conversely, for $\hat{p} > \frac{1}{2}$, the upper limit is closer to $\hat{p}$ than is the lower limit.

(b) The half-width of the interval based on the cruder approximation is $w = 1{\cdot}645\sqrt{\hat{p}(1-\hat{p})/n}$. Suppose that the first sample of 80 accounts has already been taken, so that we have $\hat{p} = 0{\cdot}275$. Then $w = 1{\cdot}645\sqrt{0{\cdot}275\times0{\cdot}725/n}$ and so $w \leq 0{\cdot}02$ if $\sqrt{n} \geq 1{\cdot}645\sqrt{0{\cdot}275\times0{\cdot}725}/0{\cdot}02$. This condition will hold if

$$n \geq \left(\frac{1{\cdot}645}{0{\cdot}02}\right)^2 \times 0{\cdot}275\times0{\cdot}725 = 1348{\cdot}8.$$

Thus, rounding up to the nearest integer we need $n \geq 1349$, so a further 1269 accounts need to be examined in addition to the 80 in the initial sample.

Notes

* (1) The alternative expression for the confidence interval given in part (a) is really too complicated to be used to estimate n in part (b). The two expressions will, in this example, give very similar numerical values for n. In any case, a very precise value for n is not usually required in practice in this sort of problem. The main objective is usually to get an approximate (perhaps conservative) idea of the sample size required to achieve the desired precision of estimation. A decision can then be made as to whether

(i) the indicated value of n is reasonable, or

(ii) the indicated value of n is too large to be feasible (which may be the case in this example), so that a lower degree of precision must be accepted, or a different method of sampling adopted, or

(iii) the indicated value of n is so small that a larger value is feasible, and hence greater precision can be achieved.

* (2) If an estimate of p is available, either from an initial sample as here, or from prior knowledge, then it can be used in the expression $z_{\alpha/2}\sqrt{\hat{p}(1-\hat{p})/n}$ for the half-width of an interval. However, it should be noted that having determined n, and taken a total sample size n, the estimate $\hat{p}$ one obtains from the whole sample will generally be different from that used to determine n. The confidence interval based on the whole sample will therefore not have exactly the required width. (It could be narrower or wider.) A conservative approach is to round up n to the nearest 10, 50 or 100, or whatever is convenient, rather than just to the nearest integer.

An alternative procedure, which guarantees that the precision will be at least that desired, is to replace p by $\frac{1}{2}$, rather than by $\hat{p}$. This is another conservative procedure since $\hat{p}(1-\hat{p})$ is maximised, and hence the interval is longest, when $\hat{p} = \frac{1}{2}$. However, the procedure is often not too conservative, unless $\hat{p}$ is very close to 0 or 1, since the function $\hat{p}(1-\hat{p})$ is fairly 'flat' in the vicinity of $\hat{p} = \frac{1}{2}$. For example, its values range only between 0·21 and 0·25 for values of $\hat{p}$ between 0·3 and 0·7. In the present example, the conservative approximation leads to $n \geq \left(\frac{1{\cdot}645}{0{\cdot}02}\right)^2 \times 0{\cdot}25$, or 1692 (rounding up to the nearest integer).

This same, conservative, approach is generally used in the fairly common situation where there is no initial sample, i.e. we wish to estimate the value of n required to achieve a given precision, but there is no estimate $\hat{p}$ of p available. Provided that p is not expected to be close to 0 or 1, the conservative approach is a reasonable one to take in these circumstances.

4C.4 Are telephone numbers random?

In an experiment to test whether final digits of telephone numbers are random, a random sample of 40 digits is selected from a large telephone directory. The hypothesis of randomness will be accepted if there are at least 17 but not more than 22 even digits in the sample, and rejected otherwise. Using a suitable approximation, find the probability of a Type I error.

A friend of yours claims that in fact more telephone numbers are odd than even. How would you check whether this is true, using a 5% test based on a random sample of size 50?

Solution

A binomial distribution is appropriate here, and if X denotes the number of even digits, we first require $\Pr(17 \le X \le 22)$ when $X \sim B(40, \frac{1}{2})$. Using a normal approximation, and remembering the continuity correction, we obtain

$$\Pr(17 \le X \le 22) \approx \Phi\left(\frac{22 + \frac{1}{2} - \mu}{\sigma}\right) - \Phi\left(\frac{17 - \frac{1}{2} - \mu}{\sigma}\right),$$

where $\mu = np = 20$ and $\sigma = \sqrt{np(1-p)} = \sqrt{10}$.

The result is, then, $\Phi(0{\cdot}79) - \Phi(-1{\cdot}11) = 0{\cdot}6517$, which is the probability of accepting the null hypothesis that $p = \frac{1}{2}$. Since a Type I error is to reject a correct null hypothesis, the probability of doing so is $1 - 0{\cdot}6517$, or $0{\cdot}3483$.

When investigating the friend's claim, the null hypothesis is, still, that $p = \frac{1}{2}$, but the alternative is now that $p < \frac{1}{2}$. Accordingly, we would reject the hypothesis only when X is unusually small for a $B(50, \frac{1}{2})$ random variable. We need to calculate the value of x such that $\Pr(X \le x) \approx 0{\cdot}05$. Using a normal approximation with continuity correction then gives

$$\Pr(X \le x) \approx \Phi\left(\frac{x + \frac{1}{2} - 25}{\sqrt{12{\cdot}5}}\right) = 0{\cdot}05,$$

and therefore

$$\frac{x + \frac{1}{2} - 25}{\sqrt{12{\cdot}5}} = -1{\cdot}645,$$

giving $x = 18{\cdot}68$. Recalling that X is in fact a discrete random variable, we would thus reject the null hypothesis that $p = \frac{1}{2}$ only if we observed 18 or fewer even digits.

Notes

(1) In the second part of the problem, the wording seems to hint that the hypothesis to test is that $p < \frac{1}{2}$. In deciding on the most appropriate null hypothesis in any problem, it is often helpful to recall that the null hypothesis can never be proved to be true; it is 'accepted' if it is consistent with the data, but that is all. Accepting a hypothesis is not the same as being convinced of its truth, since there might well be (and, indeed, always are) other hypotheses also consistent with the data.

In the present case the aim is really to see whether the hypothesis $p = \frac{1}{2}$ (strictly, $p \ge \frac{1}{2}$) is rejected as being inconsistent with the data, so that $p < \frac{1}{2}$ is the only remaining possibility. Accordingly, $p = \frac{1}{2}$ is selected as the null hypothesis, with a one-sided alternative.

(2) In the solution to the second part of the problem we rounded 18·68 down to 18 rather than to the nearest integer. This was done because the precise definition of a 5% test requires the probability of rejecting the null hypothesis when true to be *no more than* 5%. When the distributions involved are continuous, one can generally choose the critical values so that the probability is *exactly* 5%, but this is, of course, not possible with discrete data. In this problem,

the normal approximation suggests that rejecting when there are 19 or fewer even digits would give a probability of error slightly greater than the nominal 5%, and this would be unacceptable.

(In fact, use of exact calculations for $B(50, \frac{1}{2})$ rather than the normal approximation would have shown that $\Pr(X \le 18) = 0{\cdot}0325$ and $\Pr(X \le 19) = 0{\cdot}0595$. If one rejected the null hypothesis on 19 even digits, one would then have been using a significance level of about 6%, while rejecting only on 18 even digits would imply a significance level nearer 3%.)

4C.5 Smoking in public places

A sample survey was conducted to determine the opinion of a large electorate on a new policy concerning smoking in public places. A random sample of 200 men and 230 women was used, and 130 of the men and 130 women agreed with the policy. Discuss whether the two following hypotheses are tenable:

(i) that 50% of all women in the electorate agree with the policy;

(ii) that the proportion of men agreeing exceeds that of women.

Solution

It is natural, and reasonable, to assume a binomial model for this problem. In part (i), we take the number of women agreeing, X, as binomial, with $n = 230$, and unknown p (equal to p_f, say). Noting that the observed value for X is 130, we test the hypothesis $p_f = 0{\cdot}50$. The numbers involved are clearly large enough for a normal distribution approximation to be used, so we calculate

$$z = \frac{130 - \frac{1}{2} - (230\times 0{\cdot}50)}{\sqrt{230\times 0{\cdot}50\times 0{\cdot}50}} = 1{\cdot}91.$$

For a two-tailed 5% test the critical value is the ubiquitous $1{\cdot}96$, so we do not reject the hypothesis that the proportion is 50%. In the wording of the question, we find the hypothesis that 50% of women agree with the policy to be tenable.

For part (ii), we need formally to test the hypothesis $p_m = p_f$, using a natural notation. Estimates of these probabilities are given by $\hat{p}_m = \dfrac{130}{200}$ and $\hat{p}_f = \dfrac{130}{230}$. Using a normal approximation, as in part (i), the relevant distributional results are

$$\hat{p}_m \sim N\left(p_m, \frac{p_m(1-p_m)}{200}\right) \text{ and } \hat{p}_f \sim N\left(p_f, \frac{p_f(1-p_f)}{230}\right).$$

Under the null hypothesis that $p_m = p_f = p$, say, we thus obtain, approximately,

$$\hat{p}_m - \hat{p}_f \sim N\left\{0, p(1-p)\left(\frac{1}{200} + \frac{1}{230}\right)\right\}.$$

To obtain a test we thus need to estimate the common probability p; since, overall, 260 out of 430 agree, the appropriate estimate is $\dfrac{260}{430}$. We thus calculate as a test statistic

$$\frac{\dfrac{130}{200} - \dfrac{130}{230}}{\sqrt{\dfrac{260}{430}\left(1 - \dfrac{260}{430}\right)\left(\dfrac{1}{200} + \dfrac{1}{230}\right)}} = \frac{0{\cdot}0848}{0{\cdot}0473} = 1{\cdot}794.$$

This value is to be compared with percentage points of $N(0, 1)$ and, for a one-tailed test at the 5% level, the appropriate percentage point is $1{\cdot}645$. Since $1{\cdot}794 > 1{\cdot}645$ we reject, at the 5% level, the null hypothesis that the proportions are the same, and conclude that there is reasonable evidence that the proportion of men agreeing exceeds that of women.

Notes

(1) In part (i) of the question there is no mention of any alternative hypothesis. It would be excessively arbitrary to use a one-tailed test (which tail would one use?) so a two-tailed test is the only acceptable possibility.

(2) The wording of part (ii) is rather odd. Taken literally, the hypothesis stated is obviously tenable; one could hardly use the data given to refute a claim that more men than women agree with the policy. But the test performed is probably what an examiner would want; if, as here, the hypothesis of equality is rejected, it means that no hypothesis inconsistent with the one stated is tenable.

(3) Two-sample binomial tests are somewhat tricky, and while it is fairly easy to see why the distribution of

$$\frac{\hat{p}_m - \hat{p}_f}{\sqrt{p(1-p)\left(\frac{1}{200} + \frac{1}{230}\right)}}$$

is approximately $N(0,1)$, it is less straightforward to show that this holds when, in the denominator, p is replaced by a sample estimate. (One might be forgiven for thinking that the distribution might be changed from normal to Student's t, but in fact this does not happen.) The mathematical result that justifies this procedure is based on asymptotic theory, and it is reasonable to argue that with 430 observations the estimate of p will be an accurate one, so that replacing p by $\frac{260}{430}$ should have little effect.

Two-sample tests for binomial parameters can be conducted using the technique of *contingency tables*, discussed in Section 5C. The related topic of confidence intervals for problems with two binomial samples is treated in Problem 4C.6.

4C.6 Smoking habits of history and statistics teachers

In a large survey it is found that 40% of schoolteachers are smokers. Small random samples are taken of teachers of various subjects and it is found that of 20 teachers of statistics, four are smokers, whereas there are eleven smokers in a sample of 25 history teachers.

(a) Suppose that 40% of all statistics teachers smoke. Calculate, approximately, the probability that 4 or fewer are smokers in a random sample of 20 such teachers.

(b) Find an approximate 95% confidence interval for the overall proportion of statistics teachers who smoke, using the information from the sample described above.

(c) Without doing formal tests of hypotheses, use the results of (a) and (b) to discuss whether the proportion of smokers among statistics teachers is likely to be the same as that for all teachers.

(d) Find an approximate 95% confidence interval for the difference between the proportions of smokers among history teachers and among statistics teachers.

Solution

(a) The number of smokers in the sample, X, is a binomial random variable with $n = 20$ trials and probability of success $p = 0{\cdot}4$. The distribution of X is approximately normal with mean $np = 8$, and variance $np(1-p) = 4{\cdot}8$, so that

$$\Pr(X \leq 4) \approx \Pr\left(Z \leq \frac{4 + \frac{1}{2} - 8}{\sqrt{4{\cdot}8}}\right),$$

where Z is a random variable having the standard normal distribution. Thus

$$\Pr(X \leq 4) \approx \Pr\left(Z \leq \frac{-3.5}{2.19}\right)$$

$$= \Pr(Z \leq -1.60) = 0.055.$$

(b) If $\hat{p} = X/n$, then $\hat{p} \sim N(p, p(1-p)/n)$ approximately, so that an approximate confidence interval for p has end-points

$$\hat{p} \pm z_{\alpha/2}\{\hat{p}(1-\hat{p})/n\}^{\frac{1}{2}},$$

where $z_{\alpha/2}$ is an appropriate normal distribution percentage point. (The form this interval should take is discussed further in Problem 4C.3.) From the information given we obtain $\hat{p} = \frac{4}{20} = 0.2$, and $z_{\alpha/2} = 1.96$ for a 95% interval, so the limits of the interval are

$$0.2 \pm 1.96\sqrt{0.2\times0.8/20} = 0.2 \pm 0.175,$$

i.e. the confidence interval for p is $(0.025, 0.375)$.

(c) The solution to part (a) shows that, if the proportion of statistics teachers who smoke is 0·4, the probability of observing as few as four smokers in a random sample of 20 statistics teachers is rather small. (The exact probability is 0·051, as shown in Note 1.) From part (b), the approximate 95% confidence interval for p fails to include 0·4, though only just. (The improved approximation described in Note 2 gives an interval which just includes 0·4.)

Thus the results of parts (a) and (b) both suggest, although not conclusively, that the proportion of smokers among statistics teachers may be different from that for all teachers.

(d) An approximate confidence interval for the difference in proportions of smokers for the two types of teacher has end-points

$$(\hat{p}_1 - \hat{p}_2) \pm z_{\alpha/2}\left(\frac{\hat{p}_1\hat{q}_1}{n_1} + \frac{\hat{p}_2\hat{q}_2}{n_2}\right)^{\frac{1}{2}},$$

where $\hat{q}_1 = 1-\hat{p}_1$, $\hat{q}_2 = 1-\hat{p}_2$, and n_1 and n_2 are the sample sizes. For the given samples, $\hat{p}_1 = 11/25 = 0.44$, $\hat{p}_2 = 0.2$, $n_1 = 25$ and $n_2 = 20$, and for the required 95% interval $z_{\alpha/2} = 1.96$, so the interval has end-points

$$0.24 \pm 1.96\left(\frac{0.44\times0.56}{25} + \frac{0.2\times0.8}{20}\right)^{\frac{1}{2}}$$

$$= 0.24 \pm 1.96\sqrt{0.009\,86 + 0.008\,00}$$

$$= 0.24 \pm 0.262,$$

so that the 95% confidence interval for the difference between the proportions in the two groups is $(-0.022, 0.502)$.

Notes

(1) In part (a), the binomial probability could be calculated exactly as

$$\sum_{i=0}^{4}\binom{20}{i}(0.4)^i(0.6)^{20-i},$$

or looked up in published tables of the binomial cumulative distribution function. The value is in fact 0·051, which is not too different from that given by the normal approximation. Note also that a continuity correction has been incorporated into the approximate expression given in the solution. Without this correction, the probability would be approximated by

$$\Pr\left(Z \leq \frac{4-8}{\sqrt{4 \cdot 8}}\right) = \Pr(Z \leq -1 \cdot 83) = 0 \cdot 034,$$

which is substantially less accurate.

(2) The lower limit of the approximate confidence interval in (b) is rather close to zero, which is a signal that the approximation may not be particularly good. If we use the slightly better approximation described in Problem 4C.3, we find the end-points of the interval to be the roots of the quadratic equation in p

$$p^2(n + z_{\alpha/2}^2) - p(2n\hat{p} + z_{\alpha/2}^2) + n\hat{p}^2 = 0.$$

Substituting $n = 20$, $\hat{p} = 0 \cdot 2$ and $z_{\alpha/2} = 1 \cdot 96$ we find the roots to be 0·080 and 0·416. The interval is therefore (0·080, 0·416). Note that the lower limit is further from zero, but so is the upper limit; the overall width is somewhat smaller than that of the poorer approximate interval used in part (b).

(3) Although it was not asked for, the formal test of $H_0\colon p = p_0$ against $H_1\colon p \neq p_0$ (or against $H_1\colon p > p_0$ or $H_1\colon p < p_0$) is based on the test statistic

$$\frac{\hat{p} - p_0}{\sqrt{p_0 q_0 / n}},$$

which has an approximate $N(0,1)$ distribution under H_0. With $\hat{p} = 0 \cdot 2$, $p_0 = 0 \cdot 4$, $q_0 = 1 - p_0 = 0 \cdot 6$, and $n = 20$, the value of the test statistic is $-0 \cdot 2/\sqrt{0 \cdot 24/20}$, or $-1 \cdot 83$, which is identical to the value of the z-statistic without a continuity correction given in Note 1. The test statistic here can also be modified, and indeed improved, by incorporating a continuity correction, although this seems to be rarely done. Note, however, that in neither case is the test exactly equivalent to the approximate confidence interval given in the solution to (b). This is because, in the approximate confidence interval, the variance of $\hat{p}$ is estimated by $\hat{p}(1-\hat{p})/n$. However, for the test of the hypothesis $p = p_0$ the distribution of $\hat{p}$ is calculated under H_0, in which case $\mathrm{Var}(\hat{p})$ is known and is equal to $p_0(1-p_0)/n$.

* (4) When making inferences about differences between two binomial parameters p_1 and p_2, based on a normal approximation, there is not an exact equivalence between interval estimation and hypothesis testing (as in inference regarding a single binomial parameter – see Note 3 above – but unlike most inference involving means, differences of means, variances or ratios of variances). This, as in Note 3, is because of the different expressions for variances in the two situations. In interval estimation $\mathrm{Var}(\hat{p}_1 - \hat{p}_2)$ is estimated by $\hat{p}_1\hat{q}_1/n_1 + \hat{p}_2\hat{q}_2/n_2$. However, in hypothesis testing it is the variance under the null hypothesis which is required. With the standard null hypothesis $H_0\colon p_1 = p_2$, it is usual to estimate p_1 and p_2 by a common estimate $\hat{p}$, which is the overall proportion of 'successes' in the two samples combined. The variance of $\hat{p}_1 - \hat{p}_2$ is then estimated by

$$\hat{p}\hat{q}\left(\frac{1}{n_1} + \frac{1}{n_2}\right).$$

Note also that, as with differences of means, it is much more common to test $H_0\colon p_1 = p_2$ than to find a confidence interval for $p_1 - p_2$.

4C.7 Ignition problems in cars

(a) A new car model has a design fault, as a result of which 10% of production cars have a particular ignition problem. Engineers have come up with a modification which they hope will end the difficulty, and a pilot run of 30 cars is manufactured using the modified design. State and briefly justify a distribution you might consider using to model the number of cars X in this pilot run with the ignition problem.

(b) Suppose that the modification has no effect on the problem. Using your model, find $\Pr(X = 0)$ and $\Pr(X = 1)$. If in fact no cars have the problem, would the manufacturer be justified in concluding that the change has definitely improved the situation?

(c) Suppose now that the problem occurs in only 1% of cars manufactured in the modified way. Estimate the probability that in the first 500 cars there are no more than two still with the problem.

(d) Later, the manufacturer claims that as a result of the modification the problem occurs in only 0·5% of cars. An independent survey of 500 cars finds that 6 have the problem. Is the manufacturer's claim reasonable?

Solution

(a) The binomial distribution is the natural one to use in this context. Since it arises when one counts the number of 'successes' in n independent 'trials', all with the same probability p of success, it would be justified if the modified design is consistently used in the pilot run, so that the probabilities are all the same, and that the results for different cars are independent.

(b) We now assume X to have the binomial distribution with $n = 30$, and suppose that p is still 0·1. The distribution of X is thus $B(30, 0{\cdot}1)$, so that $\Pr(X = 0) = (0{\cdot}9)^{30} = 0{\cdot}0424$, and $\Pr(X = 1) = 30(0{\cdot}1)(0{\cdot}9)^{29} = 0{\cdot}1413$.

The problem faced by the manufacturer is simply one of hypothesis testing. The null hypothesis is that $p = 0{\cdot}1$ and, since the manufacturer is said to be wanting to conclude that there has been an improvement, the alternative hypothesis is one-sided. The result obtained, $X = 0$, is the most extreme possible, so the 'tail area' equivalent is simply its probability, 0·042. Since this is less than 0·05 the result is significant at the 5% level, which could be regarded as plausible evidence supporting an improvement.

(c) We now gather that the probability of trouble is reduced to 0·01. In 500 cars (assuming independence) the number X having trouble will be distributed $B(500, 0{\cdot}01)$. The required probability could, in principle, be calculated exactly from that distribution, but (in the usual notation) with n large and p very small a Poisson approximation is justified, so we can take X as Poisson $(500 \times 0{\cdot}01)$. The probability required is, thus,

$$e^{-5} + 5e^{-5} + \frac{5^2}{2!}e^{-5} = 0{\cdot}125.$$

(d) We conduct the test by comparing the observation made, 6, with the null hypothesis distribution of X, $B(500, 0{\cdot}005)$. As in part (c), the distribution is adequately approximated by the Poisson distribution with the same mean, $500 \times 0{\cdot}005 = 2{\cdot}5$. We thus calculate the probability of a result at least as extreme as 6, i.e. we find

$$\Pr(X \geq 6) = 1 - \Pr(X \leq 5) = 1 - e^{-2\cdot5}\left(1 + 2{\cdot}5 + \frac{2{\cdot}5^2}{2!} + \frac{2{\cdot}5^3}{3!} + \frac{2{\cdot}5^4}{4!} + \frac{2{\cdot}5^5}{5!}\right)$$

$$= 1 - 0{\cdot}9580 = 0{\cdot}0420.$$

The result is significant at the 5% level, providing plausible evidence to contradict the manufacturer's claim.

Notes

(1) Parts (a) and (c) of this problem are exercises in distribution theory, and could have been placed in Chapter 2. Parts (b) and (d) also require the calculation of probabilities for standard distributions, and show how such calculations appear in the context of tests of hypotheses.

(2) In part (d) we were required to test a hypothesis concerning a binomial parameter. It is not very common to find a Poisson approximation used in connection with such a test, but it is in fact the best way to proceed when the values of n and p (in the standard notation) are such that the approximation is valid.

(3) In parts (c) and (d) the Poisson distribution is used as an approximation. As mentioned in Note 2 to Problem 2A.4 such approximations are less valuable now than they once were. In fact it is not too awkward to calculate the various probabilities exactly; for example, the exact result in part (c) is 0·123. The reader will note, though, that the approximation is a very close one.

4C.8 Modifying an accident black spot

(a) At an accident black spot junction, road accidents generally occur at a rate of 5 per month, on average. After road modifications, there is 1 accident in the first month. Does this provide evidence that the modifications have reduced the risk of accidents?

(b) One member of the local district council examines the detailed records, and finds that in the two months immediately before the modifications there were 10 accidents. On the basis of this information alone, with the fact that there was 1 accident in the following month, would you conclude that there has been an improvement?

Solution

(a) To determine the significance of there being 1 accident, we need the sampling distribution of Y, the number of accidents in a month. Since accidents can be presumed to occur at random, the Poisson is the appropriate distribution. Our hypothesis is, then, that $Y \sim \text{Poisson}(5)$, and to test it against the one-sided alternative we simply require

$$\Pr(X \le 1) = e^{-5} + \frac{5e^{-5}}{1!} = 6e^{-5} = 0{\cdot}040.$$

Since this is less than 0·05 the result is significant at the 5% level, and there is therefore some reasonable evidence for an improvement in safety at the spot.

* (b) We now work on the basis of information about only three months; because of this, the mean number of accidents in a month is now unknown. The model on which analysis is based is that if X is the number of accidents in the two months before modification, and Y the number in the following month, then $X \sim \text{Poisson}(2\mu_1)$ and $Y \sim \text{Poisson}(\mu_2)$, where μ_1 and μ_2 are the monthly accident rates before and after respectively. We require to test $H_0 : \mu_1 = \mu_2$ against $H_1 : \mu_1 > \mu_2$, on the basis of observing X to be 10 and Y to be 1.

Now a two-sample test for the Poisson distribution is performed by arguing first that the total number of accidents, 11, is irrelevant to judging the hypothesis. Working conditionally on $X + Y = 11$, we obtain

$$\Pr(X = x \mid X + Y = 11) = \binom{11}{x}\left(\frac{2\mu_1}{2\mu_1 + \mu_2}\right)^x \left(\frac{\mu_2}{2\mu_1 + \mu_2}\right)^{11-x}, \quad x = 0, 1, \ldots, 11,$$

that is, conditionally on $X + Y = 11$, the distribution of X is binomial with index 11 and parameter p given by

$$p = \frac{2\mu_1}{2\mu_1 + \mu_2}.$$

(A theoretical justification of this result is provided in the solution to Problem 2A.9.) Now, under the null hypothesis that $\mu_1 = \mu_2$, $p = \frac{2}{3}$; while, under the alternative, $p > \frac{2}{3}$. The problem thus reduces to assessing an observation of 10 as coming from $B(11, \frac{2}{3})$, on a one-tailed test. The significance level is

$$\binom{11}{10}(\tfrac{2}{3})^{10}(\tfrac{1}{3}) + (\tfrac{2}{3})^{11} = 0{\cdot}0751.$$

The result is thus not significant at the 5% level, and we would not conclude on the basis of the information available that there has been an improvement.

Notes

(1) Writers are often rather casual when dealing with randomness. The wording of part (a) was devised to illustrate this, and the reader may have noticed that the solution to that part simply presumes randomness of occurrence of accidents and hence a Poisson distribution for Y. The theoretical basis for this is the *Poisson process*, a model for completely haphazard occurrences in time, discussed further in Note 1 of Problem 2A.7.

(2) There are several points in the wording of the question meriting some discussion. Trivially, one would query use of the word 'month', since months are of different lengths. In practice statisticians do not analyse data in isolation, and in this case matters of definition would obviously be cleared up in discussion with the practitioner involved.

In a similar way one would in practice query use of data for the period immediately following the modification. Warning signs are usually erected, and generally drivers approach recently modified junctions with especial care. It is customary to wait for a while until it is felt that motorists have become used to the new layout and then make a comparison as in part (b) between comparable periods. (Again, in practice, one might wish to take account of seasonal traffic loads and patterns, and perhaps weather and light conditions; in a popular summer holiday area it would be silly to compare accident rates in summer with those out of season.)

(3) The test used in part (b) is simple to apply but quite advanced in concept. The theoretical basis is given in Problem 2A.9. Intuitively, one can reason that if accidents occur at the same rate before and after, then each of the 11 accidents is twice as likely to take place in the two months before as in the one month after modification; i.e. the chance of it being before the modification is $\frac{2}{3}$. Since this holds separately for each of the 11, a binomial model results.

(4) The reader will have noticed that despite similar information – a reduction in accident rate from 5 per month to 1 in one month – being given in parts (a) and (b), the conclusions drawn in the solutions to those parts are different. This happens because the information in part (a) is in fact much more precise than that in part (b); in the latter case we have just two months' figures and could derive from them only a (fairly wide) confidence interval for μ_1, while in part (a) we are given that $\mu_1 = 5$.

We have, in effect, a particular case of a general finding that tests will be more powerful, and confidence intervals narrower, when a parameter is given than when it has to be estimated from a random sample. (A similar illustration is given by inferences about the mean of a normal distribution; when σ^2 is unknown confidence intervals are usually longer than corresponding ones when σ^2 is known: see Note 1 to Problem 4A.1.)

4D Other Problems

Finally we include some inference problems which do not fall naturally into the categories earlier in this chapter, nor into the discussion of structured data in the next chapter. The problems are basically of three types. Some deal with distributions other than the normal, binomial and Poisson, which were picked out for attention earlier; others discuss general principles, raising issues such as the value of having unbiased estimators and the meaning of confidence intervals. These are included mainly to encourage examination of basic statistical concepts, which should always be subject to constructive criticism and not just taken for granted. We also include two problems on methods of quality control. This is a topic of considerable practical importance, but often overlooked in elementary treatments of statistics.

* 4D.1 Sampling bags of apples

On a supermarket shelf there are ten $1\frac{1}{2}$kg bags of apples. Three of these bags contain 11 apples, six bags contain 10 apples, and one bag contains 9 apples. Suppose that a sample of three bags is selected randomly, without replacement, and let T be the total number of apples in the three selected bags.

(a) By considering all possible samples of three bags, obtain the sampling distribution of T.

(b) Show that the sample mean $\overline{X} = \frac{1}{3}T$ is an unbiased estimator for the mean number of apples, μ, in all ten bags.

(c) Find the variance of T, and hence the standard error of $\overline{X}$.

(d) Consider the situation where three bags are selected at random with replacement. Show that, in this case, the variance of $\overline{X}$ is 0·12.

Solution

(a) There are $\binom{10}{3} = \frac{10!}{3!\,7!} = 120$ ways of choosing three bags from ten, and under random sampling all are equally likely. Listing these possibilities gives the following table (see Note 1).

Number of bags containing x apples: $x = 9$	$x = 10$	$x = 11$	Number of ways of choosing such a sample	Value of T
1	2	0	15	29
0	3	0	20	30
1	1	1	18	30
0	2	1	45	31
1	0	2	3	31
0	1	2	18	32
0	0	3	1	33
			120	

Adding the probabilities for each value of T, we find that the sampling distribution of T has probability function $p(t) = \Pr(T = t)$ as in the table below.

t	29	30	31	32	33
$p(t)$	$\frac{15}{120}$	$\frac{38}{120}$	$\frac{48}{120}$	$\frac{18}{120}$	$\frac{1}{120}$

(b) The mean number of apples in all ten bags is

$$\mu = \frac{(1\times9) + (6\times10) + (3\times11)}{10} = 10{\cdot}2.$$

While we could calculate $E(T)$ directly, our calculations will be simplified, particularly in part (c) below, if we transform T to, say, $S = T - 30$. Then $E(T) = E(S) + 30$, and

$$E(S) = \sum_{s=-1}^{3} s\,\Pr(S = s)$$

$$= \left(-1\times\frac{15}{120}\right) + \left(0\times\frac{38}{120}\right) + \left(1\times\frac{48}{120}\right) + \left(2\times\frac{18}{120}\right) + \left(3\times\frac{1}{120}\right)$$

$$= \frac{1}{120}(-15 + 48 + 36 + 3) = \frac{72}{120} = \frac{3}{5} = 0{\cdot}6.$$

Hence $E(T) = 30{\cdot}6$, so that

$$E(\tfrac{1}{3}T) = \tfrac{1}{3}E(T) = 10{\cdot}2;$$

thus $\frac{1}{3}T$ is an unbiased estimator for μ.

(c) Since $T = S + 30$, $\text{Var}(T) = \text{Var}(S)$, and it will be much simpler to calculate $\text{Var}(S)$. Now

$$\text{Var}(S) = E(S^2) - \{E(S)\}^2$$

and

$$E(S^2) = \sum_{s=-1}^{3} s^2\,\Pr(S = s)$$

$$= \left(1\times\frac{15}{120}\right) + \left(0\times\frac{38}{120}\right) + \left(1\times\frac{48}{120}\right) + \left(4\times\frac{18}{120}\right) + \left(9\times\frac{1}{120}\right)$$

$$= \frac{1}{120}(15 + 48 + 72 + 9) = \frac{144}{120} = \frac{6}{5}.$$

It follows that

$$\text{Var}(S) = \frac{6}{5} - \left(\frac{3}{5}\right)^2 = \frac{21}{25}.$$

The variance of $\overline{X}$ is therefore

$$\text{Var}(\overline{X}) = \frac{1}{9}\text{Var}(T) = \frac{7}{75},$$

so that the standard error of $\overline{X}$ is $\sqrt{7/75} = 0{\cdot}3055$.

(d) When sampling with replacement, the variance of the sample mean is σ^2/n, where σ^2 is the variance of a single observation and n is the sample size. Let X denote a single observation. Then X has the probability function given by the table below.

x	9	10	11
$\Pr(X = x)$	1/10	6/10	3/10

We therefore obtain $E(X) = 10{\cdot}2 = \mu$, and

$$E(X^2) = \sum_{x=9}^{11} x^2 \Pr(X = x)$$

$$= \frac{1}{10}(81 + 600 + 363) = 104{\cdot}4.$$

Hence $\sigma^2 = 104{\cdot}4 - (10{\cdot}2)^2 = 104{\cdot}4 - 104{\cdot}04 = 0{\cdot}36$, so that $\sigma^2/n = 0{\cdot}12$.

Notes

(1) The entries in the table giving the sampling distribution of T are found by combinatorial arguments. For example, consider the combination (1, 1, 1). This can occur in

$$\binom{1}{1}\binom{6}{1}\binom{3}{1} = 18$$

ways. Similarly, the combination (0, 2, 1) can occur in

$$\binom{1}{0}\binom{6}{2}\binom{3}{1} = 45$$

ways, and so on.

(2) There is an alternative way of tackling part (d) which, however, takes much longer. In sampling with replacement there are $10^3 = 1000$ possible samples, each of which is equally likely to be the one chosen. Just as in the solution to part (a), one can construct a table listing the possible ways in which samples can arise, together with their probabilities of occurrence. The same reasoning as in that part of the solution can then be used to obtain Var(T) and hence Var($\overline{X}$).

(3) In the case of sampling with replacement we found the variance of $\overline{X}$ to be 0·12. The standard error of $\overline{X}$ is just the square root, i.e. 0·3465, which is larger than the value 0·3055 found for the corresponding case where sampling was done without replacement. It is natural to expect this, and indeed it will always occur, since when sampled items are not replaced more information is likely to be gained about the population.

4D.2 Two discrete random variables

A discrete random variable Y takes the values -1, 0 and 1 with probabilities $\frac{1}{2}\theta$, $1 - \theta$ and $\frac{1}{2}\theta$ respectively. Let Y_1 and Y_2 be two independent random variables, each with the same distribution as Y.

(a) List the possible values of $\{Y_1, Y_2\}$ that may arise and calculate the probability of each. Verify that your probabilities sum to unity.

(b) By calculating the value of $(Y_2 - Y_1)^2$ for each possible pair $\{Y_1, Y_2\}$, determine the sampling distribution of $(Y_2 - Y_1)^2$.

(c) Show that $X = \frac{1}{2}(Y_2 - Y_1)^2$ is an unbiased estimator for θ, and find Var(X) as a function of θ.

(d) Now suppose that $Y_1, Y_2, \ldots, Y_n$ are n independent observations on the random variable Y. Since θ is the probability that Y is not zero, a possible estimator for θ is the proportion, Z, of non-zero values among $Y_1, Y_2, \ldots, Y_n$. Write down the mean and variance of Z, and state, giving your reasons, which of X and Z you would prefer as an estimator for θ when $n = 2$.

Solution

(a) The table below gives the joint probability for each possible value taken by the pair of random variables $\{Y_1, Y_2\}$. The probabilities $\Pr(Y_1 = y_1, Y_2 = y_2)$ are given by the product $\Pr(Y_1 = y_1)\Pr(Y_2 = y_2)$, for $y_1 = -1, 0, 1$, and $y_2 = -1, 0, 1$, because Y_1 and Y_2 are independent.

		y_2		
		-1	0	1
y_1	-1	$\frac{1}{4}\theta^2$	$\frac{1}{2}\theta(1-\theta)$	$\frac{1}{4}\theta^2$
	0	$\frac{1}{2}\theta(1-\theta)$	$(1-\theta)^2$	$\frac{1}{2}\theta(1-\theta)$
	1	$\frac{1}{4}\theta^2$	$\frac{1}{2}\theta(1-\theta)$	$\frac{1}{4}\theta^2$

The sum of these nine probabilities is

$$\{4\times\tfrac{1}{4}\theta^2\} + \{4\times\tfrac{1}{2}\theta(1-\theta)\} + \{1\times(1-\theta)^2\}$$
$$= \theta^2 + 2\theta(1-\theta) + (1-\theta)^2$$
$$= \{\theta + (1-\theta)\}^2 = 1^2 = 1,$$

so that the probabilities sum to unity as required.

(b) Let $U = (Y_2 - Y_1)^2$; a table of values of U for each of the possible values of $\{Y_1, Y_2\}$ is given below.

		y_2		
		-1	0	1
y_1	-1	0	1	4
	0	1	0	1
	1	4	1	0

We see from this table that U can take just three possible values, and by combining the information from the two tables we can evaluate their probabilities. We see first that

$$\Pr(U = 0) = \tfrac{1}{4}\theta^2 + (1-\theta)^2 + \tfrac{1}{4}\theta^2$$
$$= \tfrac{1}{2}\theta^2 + (1-\theta)^2$$
$$= 1-2\theta+\tfrac{3}{2}\theta^2.$$

Similarly we obtain

$$\Pr(U = 1) = 4\{\tfrac{1}{2}\theta(1-\theta)\} = 2\theta(1-\theta),$$

and

$$\Pr(U = 4) = 2(\tfrac{1}{4}\theta^2) = \tfrac{1}{2}\theta^2.$$

The sampling distribution of U is thus as shown in the following table.

u	0	1	4
$\Pr(U = u)$	$1 - 2\theta + \frac{3}{2}\theta^2$	$2\theta(1-\theta)$	$\frac{1}{2}\theta^2$

(c) We have $X = \frac{1}{2}U$, so obtain directly $E(X) = \frac{1}{2}E(U)$ and $Var(X) = \frac{1}{4}Var(U)$. From the sampling distribution of U we find

$$E(U) = 0\times\{1 - 2\theta + \tfrac{3}{2}\theta^2\} + 1\times\{2\theta(1-\theta)\} + 4\times\{\tfrac{1}{2}\theta^2\}$$
$$= 2\theta(1 - \theta) + 2\theta^2 = 2\theta.$$

Hence $E(X) = \frac{1}{2}(2\theta) = \theta$, so X is an unbiased estimator for θ.

To obtain the variance of U (and hence that of X), we first find $E(U^2)$, as follows.

$$E(U^2) = 0^2\times\{1 - 2\theta + \tfrac{3}{2}\theta^2\} + 1^2\times\{2\theta(1-\theta)\} + 4^2\times\{\tfrac{1}{2}\theta^2\}$$
$$= 2\theta(1 - \theta) + 8\theta^2$$
$$= 2\theta + 6\theta^2.$$

We thus obtain

$$Var(U) = E(U^2) - \{E(U)\}^2$$
$$= 2\theta + 6\theta^2 - (2\theta)^2$$
$$= 2\theta + 2\theta^2 = 2\theta(1 + \theta).$$

Hence

$$Var(X) = \tfrac{1}{4}Var(U) = \tfrac{1}{2}\theta(1 + \theta).$$

(d) In a sample of n independent observations $Y_1, Y_2, \ldots, Y_n$ on the random variable Y, each of the n has the same probability θ of being non-zero. Hence V, the number of the $\{Y_i\}$ which are non-zero, is a binomial random variable with index n and parameter, i.e. probability of 'success', θ. Thus $E(V) = n\theta$, and $Var(V) = n\theta(1 - \theta)$. Hence, if $Z = V/n$, $E(Z) = \theta$ and $Var(Z) = \theta(1-\theta)/n$. Therefore Z is an unbiased estimator for θ, for any positive integer value of n, and, in particular, when $n = 2$, $Var(Z) = \frac{1}{2}\theta(1 - \theta)$.

Since $\theta > 0$, $Var(Z) = \frac{1}{2}\theta(1 - \theta) < \frac{1}{2}\theta(1 + \theta) = Var(X)$. We see then that X and Z are both unbiased estimators, but Z has the smaller variance, and on this basis Z would be preferred to X as an estimator for θ.

Notes

(1) In part (c) we found the mean and variance of X in terms of those of U, and then worked with the sampling distribution of U. We could, alternatively, have found the sampling distribution of X and then obtained $E(X)$ and $Var(X)$ directly from it. Since $X = \frac{1}{2}U$, its sampling distribution is as given in the following table.

x	0	$\frac{1}{2}$	2
$Pr(X = x)$	$1-2\theta+\frac{3}{2}\theta^2$	$2\theta(1 - \theta)$	$\frac{1}{2}\theta^2$

(2) Part (d) is somewhat open-ended, since in the statement of the problem no indication is given as to the basis of preference between X and Z. The obvious measures which one would consider using as criteria are bias and variance, and since both estimators are unbiased Z is preferred to X because of its smaller variance. But in some circumstances one may have to choose between one estimator which is unbiased but has fairly large variance and another which is slightly biased but has smaller variance, and in such cases the choice may well be a difficult one. Problem 4D.3 contains relevant material.

4D.3 Mean square error of estimators

Explain what the following statements mean:

(i) Y is a statistic,

(ii) the statistic Y is an unbiased estimator of θ.

Suppose now that a statistic Y is an unbiased estimator of a parameter θ and has variance $k\theta^2$. If one defines the mean square error MSE(X) of any estimator X of θ by

$$\text{MSE}(X) = \text{E}\{(X-\theta)^2\},$$

calculate MSE(cY), where c is some constant, and find the value of c for which MSE(cY) is a minimum.

Solution

(i) Since a statistic is a function of observed random variables, Y must be such a function.

(ii) The random variable Y has expectation θ.

We are now given that $\text{E}(Y) = \theta$, $\text{Var}(Y) = k\theta^2$, and require the mean square error of cY. From the expression given for the mean square error (MSE) of any random variable X,

$$\begin{aligned}\text{MSE}(X) &= \text{E}\{(X-\theta)^2\}\\ &= \text{E}(X^2 - 2\theta X + \theta^2)\\ &= \text{E}(X^2) - 2\theta\text{E}(X) + \theta^2\\ &= \text{Var}(X) + \{\text{E}(X)\}^2 - 2\theta\text{E}(X) + \theta^2.\end{aligned}$$

Substituting cY for X, we obtain

$$\begin{aligned}\text{MSE}(cY) &= c^2k\theta^2 + (c\theta)^2 - 2\theta.c\theta + \theta^2\\ &= \theta^2(c^2k + c^2 - 2c + 1) = m\theta^2, \text{ say.}\end{aligned}$$

To obtain the value of c for which this mean square error is a minimum, straightforward differentiation gives

$$\frac{dm}{dc} = 2kc + 2c - 2,$$

and setting this to zero gives $c = (k+1)^{-1}$. (The second derivative is positive for all values of c, so the stationary value is a minimum.)

Notes

(1) The problem centres on use of expectation and variance formulae, and helps to remind us of a variety of useful results. One of these is analogous to the result in mechanics that the moment of inertia of a body about a point is least when that point is the centre of mass. In probability and statistics this is translated into results that $\text{E}\{(X-\theta)^2\}$ and $\sum(x_i-\theta)^2$ are minimised when θ is the mean; i.e. $\theta = \text{E}(X)$ in the first case and $\theta = \bar{x}$, the sample mean, in the second. These results are related to the following useful identities:

$$\text{E}\{(X-\theta)^2\} = \text{Var}(X) + \{\text{E}(X) - \theta\}^2;$$

$$\sum_{i=1}^{n}(x_i-\theta)^2 = \sum_{i=1}^{n}(x_i-\bar{x})^2 + n(\bar{x}-\theta)^2.$$

In the case of the current problem the first of these results could have been quoted to give

$$\text{MSE}(X) = \text{Var}(X) + \{\text{E}(X) - \theta\}^2,$$

sometimes paraphrased as: ‘MSE equals variance plus bias squared’.

* (2) In the context of estimation the theory in this problem is useful in drawing attention to the possibility of using biased estimators to advantage. Elementary textbooks occasionally hint that bias is undesirable, yet in practice almost all statisticians would prefer a slightly biased estimator with small variance to an unbiased estimator with large variance; a sensible general criterion is that of mean square error.

The most straightforward example to which this theory applies is the estimation of the variance σ^2 of a normal distribution from a random sample $X_1, X_2, \ldots, X_n$. Since

$$E\{\sum_{i=1}^{n}(X_i - \overline{X})^2\} = (n-1)\sigma^2,$$

the usual estimate is s^2, the corrected sum of squares divided by $n-1$. The choice of $n-1$ as divisor may give rise to worries, but is justified, of course, on grounds of bias. But if we remove the requirement of unbiasedness other possibilities open up.

When sampling is from a normal distribution the random variable $\sum(X_i - \overline{X})^2/\sigma^2$ has a χ^2 distribution on $n-1$ degrees of freedom, which has expectation $(n-1)$ and variance $2(n-1)$. It follows that the usual estimator of σ^2 is unbiased, and has variance $2\sigma^4/(n-1)$. If we now convert to the notation of the problem, $\theta = \sigma^2$ and $k = 2/(n-1)$. The solution shows that the mean square error will be a minimum for $c = (k+1)^{-1} = (n-1)/(n+1)$. The rather surprising finding is that instead of producing an unbiased estimator by using the divisor $(n-1)$, one might consider using $(n+1)$, which would minimise mean square error. (But it is important to add that such an estimator should be used with caution; for example, it would be quite invalid to use it in the formula for a t-test, or indeed use it in place of s^2 in any other standard formula!)

4D.4 Arrivals at a petrol station

A small petrol station has a single pump. Customers arrive one at a time, their inter-arrival times independently distributed with density function

$$f(t) = \lambda e^{-\lambda t}, \quad t \geq 0,$$

for some unknown $\lambda > 0$. The times taken to serve customers are also independent and follow a normal distribution with unknown mean and variance.

(a) Discuss how you might estimate λ, and the parameters of the normal distribution, given observations on service times for n customers, together with the $n-1$ intervals between their arrivals.

(b) Show that, in a fixed time interval of length T, starting immediately after an arrival, the probability of no further arrivals is $e^{-\lambda T}$.

(c) It can be shown that the number of arrivals in any time interval of length T has a Poisson distribution with mean λT. Suggest how this result might be used to give another method for estimating λ.

(d) In a random sample of 1-hour periods the numbers of arrivals were

12, 12, 15, 18, 11, 12, 14, 17, 10, 14.

Find a 95% confidence interval for the mean number of arrivals per hour, based on a normal approximation to the Poisson distribution.

Solution

(a) For estimation of λ, we record a sample of inter-arrival times, and calculate the sample mean, $\overline{y}$, which will give a suitable estimate for the mean of the distribution, μ. Now, the relation between μ and λ is given by

$$\mu = \int_{-\infty}^{\infty} t f(t) \mathrm{d}t = \int_{0}^{\infty} t \lambda \mathrm{e}^{-\lambda t} \mathrm{d}t$$

$$= \left[-t\mathrm{e}^{-\lambda t}\right]_0^{\infty} + \int_0^{\infty} \mathrm{e}^{-\lambda t} \mathrm{d}t, \quad \text{integrating by parts,}$$

$$= \left[-\frac{1}{\lambda}\mathrm{e}^{-\lambda t}\right]_0^{\infty} = \frac{1}{\lambda}.$$

Since $\lambda = 1/\mu$ and $\bar{y}$ is an appropriate estimate for μ, we could estimate λ by $1/\bar{y}$. (There are other approaches to estimating λ based, like the present approach, on inter-arrival times, but this seems to be the most obvious.)

Estimation of the parameters of the normal distribution is relatively straightforward. A sample of service times is available, and the sample mean, $\bar{x}$, and the sample variance, s^2, can be calculated easily. These will then provide estimates for the mean and variance of the normal distribution.

(b) The probability that there are no arrivals in an interval of length T, starting immediately after an arrival takes place, is given by

$$\Pr(\text{inter-arrival time} > T) = \int_T^{\infty} f(t) \mathrm{d}t$$

$$= \int_T^{\infty} \lambda \mathrm{e}^{-\lambda t} \mathrm{d}t = \left[-\mathrm{e}^{-\lambda t}\right]_T^{\infty},$$

which is $\mathrm{e}^{-\lambda t}$, as required. (In fact the result holds for any time interval of length T, not necessarily starting immediately after an arrival.)

(c) Suppose that the number of arrivals is counted for n non-overlapping intervals each of length T. These then form a random sample of size n from a distribution with mean λT. Thus the sample mean, $\bar{x}$, will provide an estimate of λT, so that a suitable estimate of λ is $\bar{x}/T$. In fact the intervals used for counting need not all be of the same length, though they should not overlap. With unequal intervals the form of an appropriate estimate for λ will be somewhat more complicated, namely

$$\frac{\sum_{i=1}^{n} x_i}{\sum_{i=1}^{n} T_i},$$

where T_i is the length of the ith interval and x_i is the number of arrivals in that interval.

(d) Suppose now that λ is the mean number of arrivals in one hour, and that $X_1, X_2, \ldots, X_n$ form a random sample of numbers of arrivals, with sample mean $\bar{X}$. Since the mean and variance of a Poisson distribution are the same, $\mathrm{E}(\bar{X}) = \lambda$ and $\mathrm{Var}(\bar{X}) = \lambda/n$. Using a normal approximation to the distribution of $\bar{X}$ we have, approximately, $\bar{X} \sim N(\lambda, \lambda/n)$. Hence

$$\frac{\bar{X} - \lambda}{\sqrt{\lambda/n}} \sim N(0,1),$$

and so

$$\Pr\left(-z_{\alpha/2} \le \frac{\bar{X} - \lambda}{\sqrt{\lambda/n}} \le z_{\alpha/2}\right) \approx 1 - \alpha,$$

where $z_{\alpha/2}$ is, as usual, a percentage point of $N(0,1)$. Thus

$$\Pr\left(\bar{X} - z_{\alpha/2}\sqrt{\lambda/n} \le \lambda \le \bar{X} + z_{\alpha/2}\sqrt{\lambda/n}\right) \approx 1 - \alpha.$$

This, as it stands, does not give a confidence interval for λ since the variance term $\sqrt{\lambda/n}$ involves λ. However, replacing $\sqrt{\lambda/n}$ by an estimator $\sqrt{\bar{X}/n}$ gives as an approximate interval

$$\bar{X} \pm z_{\alpha/2}\sqrt{\bar{X}/n}.$$

For the given data, the sample mean is 13·5, and, for a 95% interval, $z_{\alpha/2} = 1{\cdot}96$, so the interval has end-points

$$13{\cdot}5 \pm 1{\cdot}96\sqrt{13{\cdot}5/10},$$

i.e. $13{\cdot}5 \pm 2{\cdot}28$ or $(11{\cdot}22, 15{\cdot}78)$.

Notes

* (1) The method used in the solution to part (d) is not the only possible one. It parallels the first of two methods used in the solution to Problem 4C.3 for the corresponding binomial distribution problem. In that problem an alternative solution was shown, and in the case of the Poisson distribution here we give below two other ways in which a normal approximation can be used to obtain an approximate confidence interval for λ.

One possibility is to replace $\sqrt{\lambda/n}$ by $s/\sqrt{n}$, where s^2 is the sample variance and equals 6·72 for the present data. The limits are therefore

$$13{\cdot}5 \pm 1{\cdot}96\sqrt{6{\cdot}72/10}, \quad \text{or} \quad 13{\cdot}5 \pm 1{\cdot}61,$$

i.e. the interval for λ is $(11{\cdot}89, 15{\cdot}11)$.

It would, in fact, be more appropriate here to use $t_{9,\,0{\cdot}025} = 2{\cdot}262$ rather than $z_{0{\cdot}025} = 1{\cdot}96$, since we are essentially treating our sample of 10 observations as coming from a normal distribution with mean λ and using the distribution of $\dfrac{\bar{X}-\lambda}{s/\sqrt{n}}$ to find a confidence interval for λ. In this event the interval becomes $(11{\cdot}65, 15{\cdot}35)$.

Another possibility, which gives a closer approximation, is to square the inequality

$$\left|\frac{\bar{X}-\lambda}{\sqrt{\lambda/n}}\right| \le z_{\alpha/2},$$

which holds with probability approximately $1-\alpha$, to give a quadratic inequality in λ

$$n(\bar{X}-\lambda)^2 \le z_{\alpha/2}^2\lambda,$$

or

$$n\lambda^2 - \lambda(z_{\alpha/2}^2 + 2n\bar{X}) + n\bar{X}^2 \le 0.$$

This inequality is satisfied if λ lies between the two roots λ_1 and λ_2 of the equation

$$n\lambda^2 - \lambda(z_{\alpha/2}^2 + 2n\bar{X}) + n\bar{X}^2 = 0.$$

But the probability is $1-\alpha$ that the inequality is satisfied, so that (λ_1, λ_2) gives a confidence interval for λ, with confidence coefficient $1-\alpha$. For the present example the roots of this quadratic are

$$\frac{(z_{\alpha/2}^2 + 2n\bar{X}) \pm \left[(z_{\alpha/2}^2 + 2n\bar{X})^2 - 4n^2\bar{X}^2\right]^{\frac{1}{2}}}{2n}$$

$$= \frac{273{\cdot}84 \pm [2089{\cdot}22]^{\frac{1}{2}}}{20} = 13{\cdot}69 \pm 2{\cdot}29,$$

i.e. the roots are 11·40 and 15·98, so that the 95% confidence interval is $(11{\cdot}40, 15{\cdot}98)$.

Comparing the three approximations we see that the first and the third are reasonably close, with the latter, more accurate, interval being slightly wider. The second approximation is the narrowest; this is because for the present data the sample variance is substantially smaller than would be expected, given the mean and a Poisson assumption. If the Poisson assumption is valid then this second approximate interval is probably misleadingly precise.

* (2) If a random variable X has a Poisson distribution, then $\text{Var}(X) = \text{E}(X) = \lambda$, say. Because of this, it is not at first clear that, when one has a random sample from a Poisson distribution, one should use the sample mean rather than the sample variance to estimate the parameter λ. In fact there are several cogent reasons for doing this. Some lie outside the scope of this book, but we show here that the sample mean gives the maximum likelihood estimate of λ. (See Problems 2A.6 and 2A.11 for a definition and discussion of the maximum likelihood method of estimation.)

Suppose we obtain a random sample of size n from the Poisson distribution with mean (and variance) equal to λ, and let the observations be denoted by $x_1, x_2, \ldots, x_n$. Then the probability l that these values $x_1, x_2, \ldots, x_n$ are observed is given by

$$\begin{aligned} l &= \frac{e^{-\lambda}\lambda^{x_1}}{x_1!}\frac{e^{-\lambda}\lambda^{x_2}}{x_2!}\frac{e^{-\lambda}\lambda^{x_3}}{x_3!}\cdots\frac{e^{-\lambda}\lambda^{x_n}}{x_n!} \\ &= \frac{e^{-n\lambda}\lambda^{\sum x_i}}{\prod x_i!}, \end{aligned}$$

where the sum and product are both over the range $i = 1, 2, \ldots, n$. To find the maximum likelihood estimate of λ we treat l as a function of λ, and find the value of λ at which it takes its maximum value. We could do this by differentiating l with respect to λ, but in fact it is easier, and equivalent, to differentiate $\log l$. We find

$$\log l = -n\lambda + \sum x_i \log\lambda - \log\left(\prod x_i!\right),$$

so that

$$\frac{d\log l}{d\lambda} = -n + \frac{\Sigma x_i}{\lambda},$$

and setting this to zero gives $\lambda = \Sigma x_i/n$, the sample mean $\bar{x}$. The maximum likelihood estimate of λ is thus $\bar{x}$.

(3) In this problem the arrival pattern of customers at the queue is what is termed a *Poisson process*. Such a process was defined formally in Note 1 to Problem 2A.7. We see here, from part (b), that in a Poisson process the gap between consecutive events (a continuous random variable) has an exponential distribution, while the number of events in a fixed time (a discrete random variable) has a Poisson distribution.

* 4D.5 Confidence intervals for an exponential distribution

A continuous random variable X with an exponential distribution has probability density function

$$f(x) = \begin{cases} \lambda e^{-\lambda x}, & x \geq 0, \\ 0, & \text{elsewhere.} \end{cases}$$

The mean and variance of this distribution are known to be λ^{-1} and λ^{-2} respectively.

(a) Find, in terms of λ, numbers a and b such that $\Pr(X < a) = \Pr(X > b) = 0{\cdot}05$, so that $\Pr(a \leq X \leq b) = 0{\cdot}90$. Deduce that $\Pr(0{\cdot}334X \leq \lambda^{-1} \leq 19{\cdot}496X) = 0{\cdot}90$; hence write down a 90% confidence interval for λ^{-1} based on a single observation on the random variable X.

(b) Let $\overline{X}$ denote the mean of a random sample of 81 observations on X. State the approximate distribution of $\overline{X}$, and hence show, using the approximation, that

$$\Pr(0{\cdot}821\overline{X} \le \lambda^{-1} \le 1{\cdot}278\overline{X}) = 0{\cdot}95.$$

(c) In a random sample of 81 observations on a random variable Y, the sample mean is 10·2 and the sample standard deviation is 10·6. Find a 95% confidence interval for σ^2, the variance of Y, assuming that Y has a normal distribution. Use the probability statement in part (b) to find an approximate 95% confidence interval for σ^2, if Y has in fact the same distribution as X above, rather than a normal distribution.

Solution

(a) The first of the required numbers, a, is such that

$$\int_{-\infty}^{a} f(x)\mathrm{d}x = 0{\cdot}05.$$

Now the integral is

$$\int_{0}^{a} \lambda \mathrm{e}^{-\lambda x}\mathrm{d}x = \left[-\mathrm{e}^{-\lambda x}\right]_0^a$$

$$= 1 - \mathrm{e}^{-\lambda a}.$$

Therefore $\mathrm{e}^{-\lambda a} = 1 - 0{\cdot}05 = 0{\cdot}95$, so $a = -\lambda^{-1}\log_e(0{\cdot}95) = 0{\cdot}0513\lambda^{-1}$. Similarly, b is such that $\int_b^{\infty} f(x)\mathrm{d}x = \left[-\mathrm{e}^{-\lambda x}\right]_b^{\infty} = \mathrm{e}^{-\lambda b} = 0{\cdot}05$, and therefore $b = -\lambda^{-1}\log_e(0{\cdot}05) = 2{\cdot}996\lambda^{-1}$.

Now, since $a = 0{\cdot}0513\lambda^{-1}$, we can express $\Pr(X < a)$ as $\Pr(X < 0{\cdot}0513\lambda^{-1})$ or $\Pr(\lambda^{-1} > X/0{\cdot}0513)$; we thus obtain $\Pr(X < a) = \Pr(\lambda^{-1} > X/0{\cdot}0513) = \Pr(\lambda^{-1} > 19{\cdot}496X)$. Similarly $\Pr(X > b)$ can be expressed as $\Pr(X > 2{\cdot}996\lambda^{-1})$, i.e. as $\Pr(\lambda^{-1} < X/2{\cdot}996)$, so that $\Pr(\lambda^{-1} < 0{\cdot}334X) = 0{\cdot}05$. Hence $\Pr(0{\cdot}334X \le \lambda^{-1} \le 19{\cdot}496X) = 0{\cdot}90$. The random interval $(0{\cdot}334X, 19{\cdot}496X)$ has a 90% chance of covering the parameter λ^{-1}, and is therefore a 90% confidence interval for λ^{-1}.

(b) For large samples, a sample mean $\overline{X}$ has approximately a normal distribution, regardless (almost) of the distribution of the individual observations, from the Central Limit Theorem. Here the individual observations have mean λ^{-1} and variance λ^{-2}, so that, approximately,

$$\overline{X} \sim N\left(\lambda^{-1}, \frac{\lambda^{-2}}{n}\right)$$

and

$$\frac{\overline{X} - \lambda^{-1}}{\lambda^{-1}/\sqrt{n}} \sim N(0, 1).$$

Thus

$$\Pr\left(\frac{\overline{X} - \lambda^{-1}}{\lambda^{-1}/\sqrt{n}} \le -1{\cdot}96\right) = 0{\cdot}025.$$

The inequality in the brackets is now manipulated so as to leave λ^{-1} on the left hand side, and we thus obtain

$$\Pr\left(\lambda^{-1} > \frac{\sqrt{n}\,\overline{X}}{\sqrt{n} - 1{\cdot}96}\right) = 0{\cdot}025.$$

Similarly, looking at the other tail, we find that

$$\Pr\left(\frac{\overline{X} - \lambda^{-1}}{\lambda^{-1}/\sqrt{n}} > +1{\cdot}96\right) = 0{\cdot}025$$

leads to

$$\Pr\left(\lambda^{-1} < \frac{\sqrt{n}\,\overline{X}}{\sqrt{n} + 1{\cdot}96}\right) = 0{\cdot}025.$$

Thus

$$\Pr\left(\frac{\sqrt{n}\,\overline{X}}{\sqrt{n} + 1{\cdot}96} \le \lambda^{-1} \le \frac{\sqrt{n}\,\overline{X}}{\sqrt{n} - 1{\cdot}96}\right) = 0{\cdot}95$$

and substituting $n = 81$ gives

$$\Pr\left(\frac{9\overline{X}}{10{\cdot}96} \le \lambda^{-1} \le \frac{9\overline{X}}{7{\cdot}04}\right) = 0{\cdot}90,$$

i.e. $\Pr(0{\cdot}821\overline{X} \le \lambda^{-1} \le 1{\cdot}278\overline{X}) = 0{\cdot}95$, as required.

(c) If Y has a normal distribution, and s^2 is the sample variance, then $(n-1)s^2/\sigma^2$ has a χ^2 distribution with $(n-1)$ degrees of freedom, and a confidence interval for σ^2 is given by

$$\left(\frac{(n-1)s^2}{\chi^2_{(n-1),\alpha/2}}, \frac{(n-1)s^2}{\chi^2_{(n-1),(1-\alpha/2)}}\right).$$

Now $\chi^2_{80,0{\cdot}975} = 57{\cdot}15$ and $\chi^2_{80,0{\cdot}025} = 106{\cdot}63$, so the interval is

$$\left(\frac{80\times(10{\cdot}6)^2}{106{\cdot}63}, \frac{80\times(10{\cdot}6)^2}{57{\cdot}15}\right) \quad \text{or} \quad (84{\cdot}30,\ 157{\cdot}28).$$

Finally, we are given that Y has in fact the same (exponential) distribution as X. Using the result found in part (b), we have

$$\Pr(0{\cdot}821\overline{X} \le \lambda^{-1} \le 1{\cdot}278\overline{X}) = 0{\cdot}95,$$

and simply squaring terms in the inequality (all of which are necessarily positive) gives

$$\Pr\{(0{\cdot}821\overline{X})^2 \le \lambda^{-2} \le (1{\cdot}278\overline{X})^2\} = 0{\cdot}95.$$

But λ^{-2} is the variance of Y, so $\{(0{\cdot}821\overline{X})^2, (1{\cdot}278\overline{X})^2\}$ gives a 95% confidence interval for $\mathrm{Var}(Y)$. Substituting $\overline{X} = 10{\cdot}2$ gives for this interval

$$\{(0{\cdot}821\times10{\cdot}2)^2, (1{\cdot}278\times10{\cdot}2)^2\},$$

or $(70{\cdot}13, 169{\cdot}93)$, a somewhat wider interval than that based on the normal assumption for Y.

* 4D.6 Glue sniffers

(a) If S^2 is an unbiased estimator of the variance σ^2 of some distribution, explain why S is not, in general, an unbiased estimator of the corresponding standard deviation σ.

(b) Let X_1, X_2 be the numbers of successes in two independent binomial experiments, with n_1, n_2 trials respectively and with the same probability p of success in each experiment. If $\hat{p}_1 = X_1/n_1$, $\hat{p}_2 = X_2/n_2$, and $\hat{p} = \frac{1}{2}(\hat{p}_1 + \hat{p}_2)$, show that $\hat{p}$ is an unbiased estimator for p.

Determine the range of values of the ratio n_1/n_2 for which $\mathrm{Var}(\hat{p}) < \mathrm{Var}(\hat{p}_1)$ and $\mathrm{Var}(\hat{p}) < \mathrm{Var}(\hat{p}_2)$. Show that, if $n_1 \ne n_2$, then there is an unbiased estimator for p of the form $\tilde{p} = w\hat{p}_1 + (1-w)\hat{p}_2$, where $0 < w < 1$, with smaller variance than $\hat{p}$.

(c) In order to estimate the proportion of pupils in a secondary school who were glue sniffers, it was decided to interview a random sample of pupils. On such a sensitive subject a direct question would be unlikely to elicit a truthful answer, so each pupil in the sample was asked to toss a fair coin and, without divulging the outcome to the interviewer, to answer 'Yes' or 'No' to one of two questions, depending on the outcome of the toss. If the pupil tossed a head the question to be answered was 'Is your birthday in April?'; whereas on a tail the question was 'Do you sniff glue?'. It is known that the proportion of pupils with April birthdays is 0·1. Given that the overall proportion of sampled pupils who answered 'Yes' was 0·08, estimate the proportion of pupils in the school who sniff glue.

Solution

(a) In general, if X is a random variable, and $g(X)$ is a function of X, then $\mathrm{E}\{g(X)\} \neq g\{\mathrm{E}(X)\}$, except when $g(X)$ is a linear function of X or, trivially, when X is a constant. Thus, if X is an unbiased estimator of θ, then $g(X)$ will not, in general, be an unbiased estimator of $g(\theta)$. This part of the problem is concerned with a special case of this general result, where $X = S^2$ and $g(X) = \sqrt{X}$.

In this special case the result can be proved directly without much trouble. Consider the variance of S:

$$\mathrm{Var}(S) = \mathrm{E}[\{S - \mathrm{E}(S)\}^2].$$

This will be greater than zero unless S is a constant. But

$$\mathrm{Var}(S) = \mathrm{E}(S^2) - \{\mathrm{E}(S)\}^2,$$

so that

$$\{\mathrm{E}(S)\}^2 < \mathrm{E}(S^2) = \sigma^2;$$

$$\text{i.e.} \quad \mathrm{E}(S) < \sigma.$$

(b) Quoting results for binomial distributions, $\mathrm{E}(X_1) = n_1 p$ and $\mathrm{E}(X_2) = n_2 p$. We thus find that

$$\mathrm{E}(\hat{p}_1) = \mathrm{E}(\hat{p}_2) = p,$$

and hence that

$$\mathrm{E}(\hat{p}) = \tfrac{1}{2}\{\mathrm{E}(\hat{p}_1) + \mathrm{E}(\hat{p}_2)\} = \tfrac{1}{2}(p + p) = p,$$

so that $\hat{p}$ is an unbiased estimator of p. Further, from standard results, $\mathrm{Var}(X_1) = n_1 p(1-p)$ and $\mathrm{Var}(X_2) = n_2 p(1-p)$, so $\mathrm{Var}(\hat{p}_1) = p(1-p)/n_1$ and $\mathrm{Var}(\hat{p}_2) = p(1-p)/n_2$. Because the two experiments are independent, so are $\hat{p}_1$ and $\hat{p}_2$. By the additive properties of variance for independent random variables we have

$$\mathrm{Var}(\hat{p}) = \tfrac{1}{4}\{\mathrm{Var}(\hat{p}_1) + \mathrm{Var}(\hat{p}_2)\} = \tfrac{1}{4}p(1-p)\left(\frac{1}{n_1} + \frac{1}{n_2}\right).$$

Hence the required condition $\mathrm{Var}(\hat{p}) < \mathrm{Var}(\hat{p}_1)$ will be satisfied if

$$\tfrac{1}{4}p(1-p)\left(\frac{1}{n_1} + \frac{1}{n_2}\right) < \frac{p(1-p)}{n_1},$$

i.e. if

$$\frac{n_1 + n_2}{n_1 n_2} < \frac{4}{n_1},$$

or

$$n_1 + n_2 < 4n_2.$$

This condition reduces to $n_1 < 3n_2$, or $n_1/n_2 < 3$. Similarly, $\text{Var}(\hat{p}) < \text{Var}(\hat{p}_2)$ if $n_2/n_1 < 3$, and so the two conditions quoted both hold if

$$\frac{1}{3} < \frac{n_1}{n_2} < 3.$$

To show that there is an unbiased estimator of p of the form given with a smaller variance than $\hat{p}$, consider $\tilde{p} = w\hat{p}_1 + (1-w)\hat{p}_2$, where $0 \le w \le 1$. Then

$$\begin{aligned} \text{E}(\tilde{p}) &= w\text{E}(\hat{p}_1) + (1-w)\text{E}(\hat{p}_2) \\ &= wp + (1-w)p = p, \end{aligned}$$

so that $\tilde{p}$ is seen to be unbiased. For its variance we have

$$\text{Var}(\tilde{p}) = w^2\text{Var}(\hat{p}_1) + (1-w)^2\text{Var}(\hat{p}_2),$$

since $\hat{p}_1$ and $\hat{p}_2$ are independent. Thus

$$\begin{aligned} \text{Var}(\tilde{p}) &= w^2\frac{p(1-p)}{n_1} + (1-w)^2\frac{p(1-p)}{n_2}, \\ &= p(1-p)\left\{\frac{w^2}{n_1} + \frac{(1-w)^2}{n_2}\right\}, \\ &= \frac{p(1-p)}{n_1 n_2}[w^2(n_1+n_2) - 2wn_1 + n_1], \\ &= \frac{p(1-p)}{n_1 n_2}\left[(n_1+n_2)\left(w - \frac{n_1}{n_1+n_2}\right)^2 + n_1 - \frac{n_1^2}{n_1+n_2}\right]. \end{aligned}$$

We see from this that $\text{Var}(\tilde{p})$ depends on w only through the first term in the square bracket, which contains a squared component and is therefore at a minimum when $w = n_1/(n_1+n_2)$; for any other value of w, $\text{Var}(\tilde{p})$ will be greater. But the estimator $\hat{p}$ corresponds to $w = \frac{1}{2}$, so that $\text{Var}(\tilde{p}) < \text{Var}(\hat{p})$ unless $w = \frac{1}{2}$, i.e. unless $n_1 = n_2$.

(c) The probability that a pupil returns a 'Yes' answer is

$$\Pr(\text{'Yes'} \mid A_1)\Pr(A_1) + \Pr(\text{'Yes'} \mid A_2)\Pr(A_2),$$

where A_1 and A_2 are respectively the events that the first and second questions are answered. Let the proportion of pupils who sniff glue be p. Then

$$\begin{aligned} \Pr(\text{'Yes'}) &= \Pr(\text{birthday in April})\Pr(\text{Head}) + \Pr(\text{sniffs glue})\Pr(\text{Tail}) \\ &= \frac{1}{10}\times\frac{1}{2} + p\times\frac{1}{2} = \frac{p}{2} + \frac{1}{20}. \end{aligned}$$

If we equate this expression to the observed proportion answering 'Yes', 0·08, and solve the equation for p, the result is an intuitively reasonable estimator of p. (In fact it is unbiased, and has other desirable properties.) We thus solve

$$\begin{aligned} 0{\cdot}08 &= 0{\cdot}5\hat{p} + 0{\cdot}05, \\ \text{i.e. } \hat{p} &= 2\times(0{\cdot}08 - 0{\cdot}05) \\ &= 0{\cdot}06. \end{aligned}$$

Thus the proportion of glue sniffers at the school is estimated to be 0·06, or 6%.

Notes

(1) All three parts of this problem are difficult; in particular, parts (a) and (c) are non-standard and somewhat open-ended. However, questions similar to all three parts have appeared as parts of (separate) public examination questions in the U.K.

(2) There is an alternative derivation of the value of w which, in part (b), minimises Var($\tilde{p}$). This is in some respects a more natural approach, since to many people differentiation is the most straightforward way of identifying a minimum. If we differentiate the expression for Var($\tilde{p}$) we obtain

$$\frac{d}{dw}\text{Var}(\tilde{p}) = \frac{d}{dw}\left\{p(1-p)\left(\frac{w^2}{n_1} + \frac{(1-w)^2}{n_2}\right)\right\},$$

$$= p(1-p)\left\{\frac{2w}{n_1} - \frac{2(1-w)}{n_2}\right\}.$$

Equating this to zero gives $2w/n_1 = 2(1-w)/n_2$, and we thus obtain the equation $n_2 w = n_1 - n_1 w$, whose solution is, as expected, $w = n_1/(n_1 + n_2)$. Further, we can differentiate again to obtain

$$\frac{d^2}{dw^2}\text{Var}(\tilde{p}) = 2p(1-p)\left(\frac{1}{n_1} + \frac{1}{n_2}\right)$$

which is necessarily positive for valid values of p, n_1 and n_2. The stationary value for w is therefore a minimum, as required.

(3) The final part of the question is open-ended in several respects, but the solution given is the most natural one. A less formal way of looking at the problem is to argue that since the two questions have the same chance of being answered, the overall probability of a 'Yes' answer is a simple average of the two individual probabilities. Setting this average equal to the observed proportion giving a 'Yes' answer, and solving, gives

$$0{\cdot}08 = \tfrac{1}{2}(0{\cdot}1 + \hat{p}),$$

which gives $\hat{p} = 0{\cdot}06$, as before.

(4) The interviewing technique discussed in part (c) is known as the *randomised response technique*. Another version of the technique is described in Problem 1B.5.

4D.7 Quality control for bolts

A machine produces bolts, and for each bolt there is a probability p of it being defective, results for different bolts being independent. A large batch of the machine's production is inspected by a customer in order to determine whether the batch should be purchased. In the inspection, 10 bolts are selected at random and examined: if none is defective the batch is accepted, and if three or more are defective it is rejected. If only one or two are defective, a further sample of 10 is selected, and the batch is accepted if, in total, there are no more than two defectives.

(a) What is the probability of a decision being made at the first stage?

(b) Find the probability that the batch is accepted (i) if $p = 0{\cdot}05$, (ii) if $p = 0{\cdot}15$.

Solution

(a) A decision is made at the first stage unless the number of defectives is 1 or 2. The required probability is thus

$$1 - \binom{10}{1}p(1-p)^9 - \binom{10}{2}p^2(1-p)^8 = 1 - 5p(2+9p)(1-p)^8.$$

(b) It is easiest to work algebraically in the first instance, and to let A denote overall acceptance, and B_0, B_1, B_2 and B_3 denote, respectively, 0, 1, 2 and 3 *or more* defectives at the first stage. (We denote the corresponding events for the *second* stage by B_0^*, B_1^*, B_2^* and B_3^*.) Clearly B_0, B_1, B_2 and B_3 are mutually exclusive and exhaustive, so the conditions for the law of total probability are satisfied, and we have

$$\Pr(A) = \sum_{i=0}^{3} \Pr(B_i)\Pr(A \mid B_i).$$

Further, $\Pr(B_0) = (1-p)^{10}$, $\Pr(B_1) = 10p(1-p)^9$, $\Pr(B_2) = 45p^2(1-p)^8$, and $\Pr(B_3)$ could if needed be found by subtraction. The conditional probabilities are quite easily found. Trivially $\Pr(A \mid B_0) = 1$, $\Pr(A \mid B_3) = 0$. If B_1 occurs, then acceptance follows if the second sample contains no more than 1 defective; i.e.

$$\Pr(A \mid B_1) = \Pr(B_0^*) + \Pr(B_1^*).$$

Similarly

$$\Pr(A \mid B_2) = \Pr(B_0^*).$$

We thus obtain

$$\begin{aligned} \Pr(A) &= \Pr(B_0) + \Pr(B_1)\{\Pr(B_0^*) + \Pr(B_1^*)\} + \Pr(B_2)\Pr(B_0^*) \\ &= (1-p)^{10}\{1 + 10p(1-p)^9 + 145p^2(1-p)^8\}, \end{aligned}$$

since $\Pr(B_i^*) = \Pr(B_i)$, $i = 0, 1, 2, 3$. Substituting $p = 0{\cdot}05$ gives $\Pr(A) = 0{\cdot}93$, while when $p = 0{\cdot}15$ the probability becomes $0{\cdot}44$.

Notes

(1) While the problem contains relatively straightforward numbers the set-up described is quite realistic, in that a double sampling scheme of this type is quite frequently employed in quality control work. (Often the sample sizes will be somewhat larger, and in many cases one will be interested in rather smaller values for p.) The probability of batch acceptance, treated as a function of the proportion p of defectives, is termed the *operating characteristic* or O.C. of the procedure, and in practice an important aim is to achieve a satisfactory O.C. without excessive cost. For example, in practice one might wish to compare the O.C. for the current sampling scheme with that for a scheme which samples, say, 8 at the first stage, with a further 15 if a second stage is needed. Considerable research effort has gone into devising suitable schemes for all situations, and tables containing details of these schemes are published by various bodies concerned with manufacturing standards, for example the British Standards Institution.

(2) In the current case, and indeed in most elementary problems involving ideas of quality control, items sampled are classified only as defective or otherwise; the resulting sampling schemes are described as *sampling by attributes*. In many practical cases items will be acceptable if a measurement lies within certain limits, often called *tolerance limits*, and if such a measurement is made it offers opportunities for using extra information in a sampling scheme. Such schemes are described as *sampling by variables*, and can offer significant savings in effort over the simpler attribute sampling schemes.

4D.8 A control chart for mass-produced articles

Articles are mass-produced to a specified width of 0·15 cm. In order to check that the production process is running satisfactorily, random samples of five articles are taken at regular time intervals, and the widths measured. The following are the means and ranges of the widths in 20 consecutive samples, where the widths are measured from the specified value in units of 0·0001 cm.

Sample number	Sample mean	Sample range	Sample number	Sample mean	Sample range
1	1·0	32	11	−1·6	18
2	2·4	21	12	−2·2	14
3	−2·4	19	13	1·6	6
4	3·2	6	14	0·4	15
5	1·4	11	15	2·0	18
6	1·2	17	16	3·0	16
7	2·6	13	17	1·6	18
8	0·4	9	18	−7·6	11
9	−2·2	14	19	−5·4	15
10	−3·8	8	20	−4·2	8

Using these results, convert the average range into an estimate of the standard deviation of the widths of the articles in the population of manufactured articles. (If the range of a random sample of size 5 from a normal population is r, then an estimate of the standard deviation is given by multiplying r by 0·4299.)

Hence draw a control chart for means, showing both 2·5% control limits (warning limits) and 0·1% control limits (action limits). Describe how this chart could be used to determine whether or not the manufacturing process is under control. What would your conclusion be for the data provided?

Solution

From the values of the 20 sample ranges given, the mean is 289/20 = 14·45, in working units, or $14{\cdot}45\times10^{-4}$ cm. The average range of samples of size 5 (assuming that they are taken from a normal population) provides an estimate of the population standard deviation equal to $0{\cdot}4299\times14{\cdot}45\times10^{-4} = 6{\cdot}2121\times10^{-4}$ cm. Hence an estimate of the standard error of the mean of a sample of size 5 is $6{\cdot}2121/\sqrt{5} = 2{\cdot}778$, in the working units.

From tables of the standardised normal distribution, $\Phi(2{\cdot}242) = 0{\cdot}9875$, so that 2·5% control limits are at a distance $2{\cdot}242\times2{\cdot}778 = 6{\cdot}23$ either side of zero. (Recall that the measurements quoted are the discrepancies between the actual widths and the specified value, 0·15 cm, in units of 0·0001 cm.)

Similarly, 0·1% control limits are obtained by scanning tables of the standardised normal distribution to find the value x such that $\Phi(x) = 0{\cdot}9995$; we obtain $x = 3{\cdot}290$, and conclude that the control limits are $3{\cdot}290\times2{\cdot}778 = 9{\cdot}14$ either side of zero.

The control chart for means is therefore as shown in Figure 4.1. Inspecting the diagram, we see that just one mean (from sample 18) falls outside the 2·5% (warning) limits, and none outside the 0·1% (action) limits. Now, even if the process were under control, we would expect, on average, one in 40 of the sample means to fall outside the 2·5% limits, and here one out of 20 so far has done so. This, therefore, provides some warning that the process might be getting out of control, but the evidence is far from conclusive, and no immediate action would be taken.

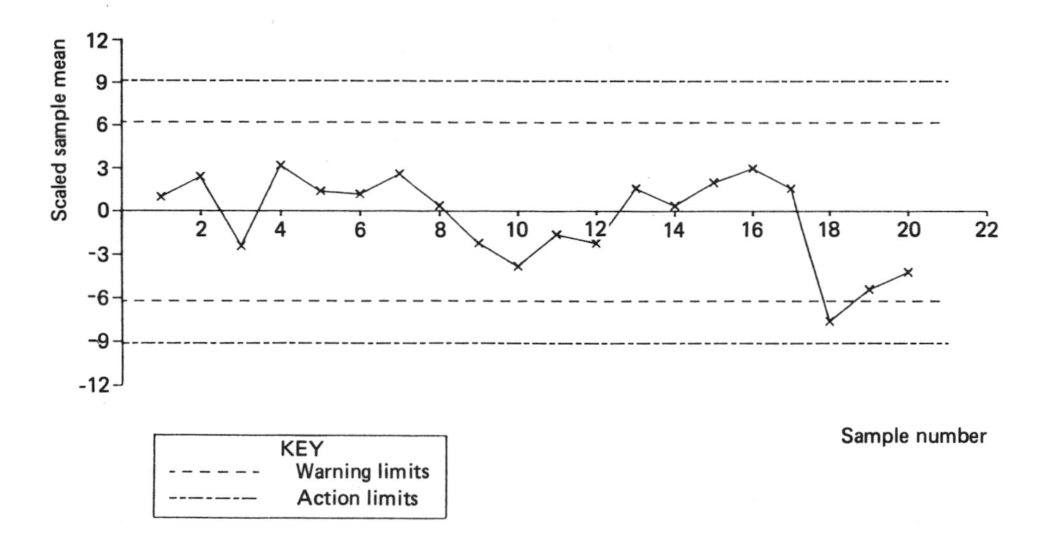

Figure 4.1 Control chart for Problem 4D.8

One should also examine the pattern of the points in Figure 4.1 over time to check, for example, whether the means have an upward or downward trend, rather than fluctuating randomly. Such trends, or other obvious patterns, do not seem to be discernible here.

Notes

(1) Multiplying factors for converting sample ranges to estimates of population standard deviations, for various sample sizes, are tabulated in some books of statistical tables. One such source, from which the value 0·4299 was obtained, is *Elementary Statistics Tables*, by H. R. Neave.

(2) A similar control chart could also be constructed for the ranges, to provide an additional check on the process. If this is done for the data provided, the conclusions are much the same as for the mean.

(3) We note at the end of the solution that the pattern of the points should be examined to test for a trend or some other regular (perhaps cyclic) variation. The techniques of time series analysis provide us with appropriate tests; in particular the statistic S described in Problem 5D.2 can be used to test for a trend here.

(4) In Figure 4.1 the points representing the sample means are joined by straight lines. This is an optional feature of a control chart; generally, including these lines can make any trend or cyclic pattern easier to see. But we have seen charts, especially when the number of points plotted is large, when joining them by lines obscures details and is thus a hindrance.

5 Analysis of Structured Data

As in Chapter 4, the problems in this chapter are all concerned with inference from sample data. In that chapter we discussed problems in which one or two random samples had been selected from some distribution, usually the normal, binomial or Poisson, and the inferences required concerned the parameter or parameters of the distribution concerned. But the data in these problems had very little structure, and had the values been presented in a different order the conclusions would have remained the same. (In the two-sample problems the allocation of observations to samples would, of course, have to remain the same!) By contrast, the data in each of the sections of this chapter present a well-defined structure, and the problems are thus slightly more advanced than most of those in Chapter 4. There is, however, little in common between the various sections of this chapter, and we therefore keep this general introduction to a minimum, preferring to introduce the individual sections at greater length.

5A Regression and Correlation

People often confuse the techniques of regression and correlation. Mathematically, they are rather similar, and either could be applied baldly to many data sets. Both apply to situations in which measurements are recorded on two variables, x and y say. But their essential difference lies in the process of specifying an appropriate mathematical model for the data.

If the aim is simply to assess the strength of the relationship between x and y, and both can be regarded as random variables, then correlation is appropriate; by contrast, regression is used when x and y are *not* to be treated symmetrically, perhaps when only one is random and the other is under an experimenter's control. Of course, little is clear-cut in statistics, and there are circumstances in which both techniques can legitimately be used. But it must be said that misuse is very frequent (see, for example, Problem 5A.4), and for some reason this is particularly true of correlation.

The majority of examination questions on correlation, at the level we are considering, seek very little in the way of understanding on the part of the candidate; all that is required is familiarity with computational techniques and a few mechanical procedures for hypothesis testing and interval estimation. This reflects two facts. One is that a correlation coefficient is very frequently, like a variance, a 'nuisance parameter' which has to be estimated but is not itself of particular interest. The other is that situations where correlation coefficients *are* of interest to the statistician typically involve, rather than a single sample from a bivariate distribution, one or several samples from multivariate distributions. The problems on correlation we have included here represent the 'nursery slopes' of the study of correlation, in the same sense as the z-test for a normal mean, appropriate to the somewhat unrealistic situation when the variance is known, paves the way for the more useful t-test.

In many respects the same is true of the regression problems to be found here; we confine ourselves to cases in which there is a single explanatory (or 'independent') variable, although in statistical practice one is usually faced with many possible variables, the aim being to determine which, if any, help to explain the observed variability in the dependent variable y. For example, as this note was being written, we received a report of a survey into the annual income of statisticians in the United Kingdom. In that survey annual income was the dependent variable, and sex, age, location (in London or elsewhere), number of jobs held, period in current job, and qualifications were, amongst others, explanatory variables.

We noted above that the two techniques are similar mathematically. Since they share the same basic calculations, it is natural that we should present these using a common notation. We suppose that the data available consist of n pairs of observations, $(x_1, y_1), (x_2, y_2), \ldots, (x_n, y_n)$. The three major quantities needed are the *corrected sums of squares and products,* which we denote by S_{xx}, S_{yy} and S_{xy}. The corrected sums of squares S_{xx} and S_{yy} are the familiar quantities which appear when we calculate a sample variance. The former is *defined* as

$$S_{xx} = \sum_{i=1}^{n}(x_i - \bar{x})^2,$$

but is almost always *calculated* as

$$S_{xx} = \sum_{i=1}^{n}x_i^2 - \frac{1}{n}\left(\sum_{i=1}^{n}x_i\right)^2.$$

The expressions for S_{yy} are similar, while the corrected sum of products is

$$S_{xy} = \sum_{i=1}^{n}(x_i - \bar{x})(y_i - \bar{y})$$

$$= \sum_{i=1}^{n}x_i y_i - \frac{1}{n}\left(\sum_{i=1}^{n}x_i\right)\left(\sum_{i=1}^{n}y_i\right),$$

the former being the definition and the latter being more suitable for calculation.

It follows that the basic calculations required from a sample of pairs (x_i, y_i), $i = 1, 2, \ldots, n$, are the sums Σx_i and Σy_i, the sums of squares Σx_i^2 and Σy_i^2 and the sum of products $\Sigma x_i y_i$, where in each case the sum is over the range from $i = 1$ to n.

In the first few problems below we give the calculations in detail, including explicit limits on summations, and with each step given. But such detail does get tedious, and in later problems we omit some intermediate details of calculations and streamline notation somewhat; thus, for example, $\sum_{i=1}^{n} x_i y_i$ can in many circumstances be shortened to Σxy without loss of clarity.

5A.1 The numbers of pensioners

The data below show, for the years given, the numbers (to the nearest thousand) receiving U.K. retirement pensions.

Year	1966	1971	1975	1976	1977	1978	1979
No. of retirement pensions (thousands)	6679	7677	8321	8510	8637	8785	8937

Plot the data and fit an appropriate least-squares regression line, drawing the line on your graph. Is there convincing evidence that the number of pensioners is increasing?

Use your results to estimate the expected number of pensioners in 1979, 1981, 1985 and 2000, and comment on your findings.

Solution

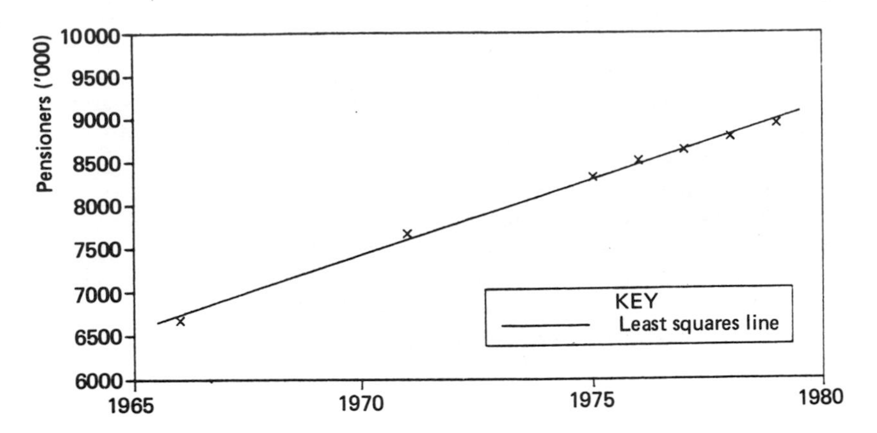

Figure 5.1 Data on retirement pensioners, from Problem 5A.1

The required plot is shown in Figure 5.1. To calculate the least-squares regression line for numbers like the ones above it is simplest to use some form of coding; an arbitrary but fairly sensible coding scheme subtracts 1970 from the year and 8000 from the number of pensioners (itself in thousands), dividing the result by 100. The data are then as follows.

Year	(x)	−4	1	5	6	7	8	9
Pensioners	(y)	−13·21	−3·23	3·21	5·10	6·37	7·85	9·37

The following basic calculations start the analysis:

$$\sum_{i=1}^{7} x_i = 32{\cdot}0, \qquad \sum_{i=1}^{7} y_i = 15{\cdot}46,$$

$$\sum_{i=1}^{7} x_i^2 = 272{\cdot}00, \qquad \sum_{i=1}^{7} y_i^2 = 411{\cdot}247,$$

$$\sum_{i=1}^{7} x_i y_i = 287{\cdot}98.$$

We next obtain the corrected sums of squares and products:

$$S_{xx} = 272{\cdot}00 - \frac{32{\cdot}0^2}{7} = 125{\cdot}714,$$

$$S_{yy} = 411{\cdot}247 - \frac{15{\cdot}46^2}{7} = 377{\cdot}103,$$

$$S_{xy} = 287{\cdot}98 - \frac{32{\cdot}0 \times 15{\cdot}46}{7} = 217{\cdot}306.$$

The estimated regression line has gradient $\hat{\beta} = \dfrac{217{\cdot}306}{125{\cdot}714} = 1{\cdot}729$, and its intercept is given by $\hat{\alpha} = \dfrac{15{\cdot}46 - (1{\cdot}729 \times 32{\cdot}0)}{7} = -5{\cdot}693$. The line $y = -5{\cdot}693 + 1{\cdot}729x$ is shown in Figure 5.1, together with the data given in the problem.

We are asked to consider evidence that the number of pensioners is increasing, and to do this we examine the null hypothesis that it is stationary; that is, that the slope β of the true regression line is zero. Under this null hypothesis, $E(\hat{\beta}) = 0$ and $\mathrm{Var}(\hat{\beta}) = \sigma^2/S_{xx}$, where σ^2 is the error variance; we note that σ^2 is estimated by

$$s^2 = \frac{1}{n-2}\left(S_{yy} - \frac{S_{xy}^2}{S_{xx}}\right) = 0{\cdot}2950.$$

We thus compare $\dfrac{1{\cdot}729}{\sqrt{0{\cdot}2950/125{\cdot}7}} = 35{\cdot}68$ with critical values of the t-distribution on 5 degrees of freedom; the 0·1% critical value, for a one-tailed test, is 5·89, and the evidence that the number of pensioners is increasing is therefore compelling.

The regression line found is

$$y = -5{\cdot}693 + 1{\cdot}729x,$$

with both x and y in coded form. Predictions for y are required for values of x coded as 9, 11, 15 and 30. Substituting in the regression equation gives, respectively, 9·864, 13·321, 20·235 and 46·164 for coded values of y and 8986·4, 9332·1, 10 023·5, 12 616·4 for the values in the original scale.

A natural comment to make is that the results show the dangers of extrapolation; it might be reasonable to try to predict for a year or two, but predictions for further ahead are increasingly implausible. One would couple these findings with the curvature evident in the plot; it appears from the latter that the growth in the number of pensioners is not linear, but started to tail away before 1979. Thus all the predictions seem likely to exaggerate the actual number of pensioners, the later ones grossly so.

Notes

(1) In many cases there can arise legitimate doubts about whether correlation or regression is the more appropriate technique. When one of the variables is time, however, no such doubt is possible. Correlation is valid when both measurements are on random variables, and is thus inappropriate here.

(2) Most statisticians would feel that the relationship between y and x is not adequately described by a straight line with random scatter, but that a curved relationship is indicated. A common way of proceeding is to test adequacy of the linear model

$$y = \alpha + \beta x + \epsilon,$$

where ϵ represents a random error term, by comparing the fit of this model with that of the quadratic regression model

$$y = \alpha + \beta x + \gamma x^2 + \epsilon.$$

We do not have the space here to discuss fully how to fit the latter model, but readers will see that the best-fitting parabola

$$y = -5{\cdot}217 + 1{\cdot}872x - 0{\cdot}0291x^2$$

gives a very much closer fit to the data. In the table overleaf, we show, for each value of x, the observed value of y, together with the predicted, or fitted, values given by the two equations, linear and quadratic. (All quantities are coded as in the solution.) A simple way of assessing the fit of an equation is to calculate the differences between the observed and fitted values; these discrepancies, usually termed *residuals*, are also given in the table and are plotted in Figure 5.2.

The plot of the residuals from the linear equation has a distinctly curved shape, and in itself indicates that such a model is inadequate. By contrast, the residuals from the quadratic model appear haphazard. (Note 2 to Problem 5D.2 shows how one could approach the testing of these residuals for randomness, although a satisfactory test would require far more observations than have been given here.)

x	Observed y	Fitted Values (Linear)	Fitted Values (Quadratic)	Residuals (Linear)	Residuals (Quadratic)
−4	−13·210	−12·608	−13·170	−0·602	−0·040
1	−3·230	−3·965	−3·375	0·735	0·145
5	3·210	2·949	3·414	0·261	−0·204
6	5·100	4·678	4·966	0·422	0·134
7	6·370	6·407	6·459	−0·037	−0·089
8	7·850	8·135	7·894	−0·285	−0·044
9	9·370	9·864	9·272	−0·494	0·098

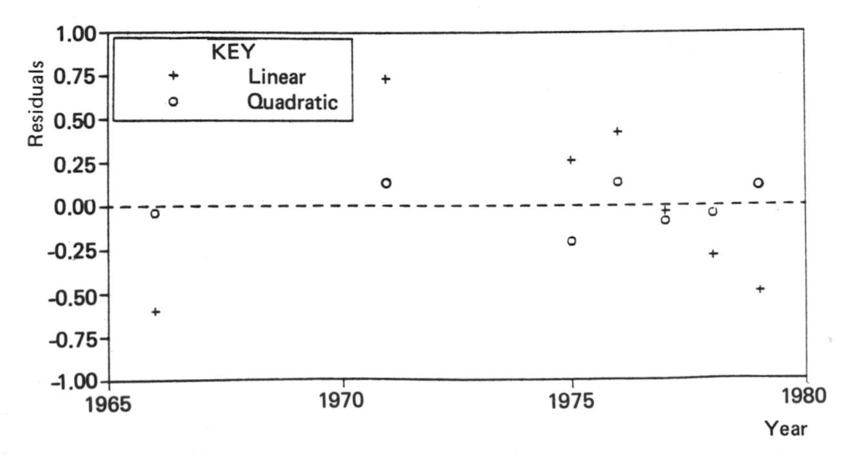

Figure 5.2 Residuals from linear and quadratic models of Problem 5A.1

(3) In the solution we pointed out the dangers of extrapolation from the straight-line model. Despite the improved fit of the quadratic model, extrapolation from this will also be dangerous, though a prediction of $y = 13{\cdot}316$ (corresponding to 9631·3 in the original scale) for 1985 is more plausible than the linear prediction of 20·235 (corresponding to about 10 024, i.e. over 10 million pensioners).

(4) The formulae used in the solution for $\hat{\alpha}$ and $\hat{\beta}$, $\hat{\beta} = S_{xy}/S_{xx}$ and $\hat{\alpha} = \bar{y} - \hat{\beta}\bar{x}$, are obtained using the *method of least squares*. The principle employed here is that, when fitting a line $y = \alpha + \beta x$ to a set of points $(x_1, y_1), \ldots, (x_n, y_n)$, we wish the discrepancies $|y_i - \alpha - \beta x_i|$ to be, in some sense, as small as possible. A measure of the quality of the fit of the line to the points is given by

$$\sum_{i=1}^{n} (y_i - \alpha - \beta x_i)^2,$$

and the best-fitting line then has slope and intercept given by the expressions at the start of this note. Since the estimates are found by minimising the sum of squares of the discrepancies, they are naturally known as least squares estimates. The method is widely used in regression and analysis of variance problems; it was used in Note 2 to fit the second-degree model by choosing α, β and γ to minimise the expression

$$\sum_{i=1}^{n} (y_i - \alpha - \beta x_i - \gamma x_i^2)^2.$$

5A.2 Graduate unemployment

It has been suggested that the unemployment rate for graduates of U.K. universities may depend on the proportion of Arts students in those universities. A random sample of 9 universities (excluding any with medical students) gave the following results. Do you conclude that the suggestion is justified?

Percentage of Arts students	Percentage of graduates unemployed
34·81	20·6
32·05	13·0
65·37	24·4
64·09	19·0
75·61	25·5
44·82	16·7
53·96	16·7
64·00	24·7
69·35	21·5

Solution

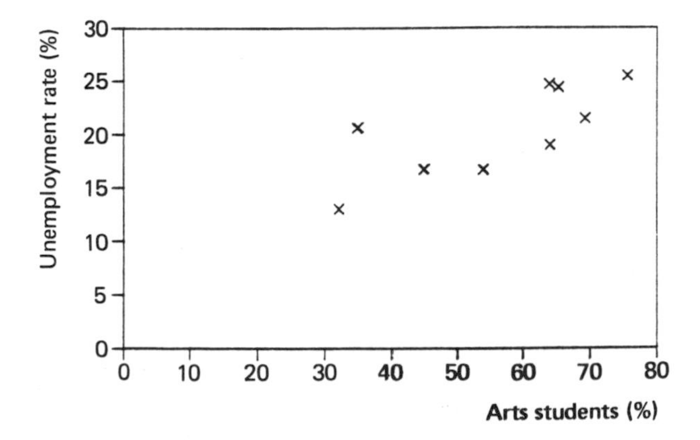

Figure 5.3 Unemployment percentage data of Problem 5A.2

Figure 5.3 shows a plot of the data, in which there is a clear indication of a relationship between the two variables. Since the suggestion that unemployment *depends* on subject balance within a university implies that we have one explanatory and one dependent variable, regression is the appropriate technique to use. We employ a natural notation, with arts percentage denoted by x and percentage unemployed by y. Calculation of sums, sums of squares and sum of products gives the following: $n = 9$, $\Sigma x = 504{\cdot}06$, $\Sigma y = 182{\cdot}1$, $\Sigma x^2 = 30\,162{\cdot}5$, $\Sigma y^2 = 3830{\cdot}09$, $\Sigma xy = 10\,596{\cdot}0$, where the sums are taken over the nine universities.

We can now proceed to estimate the gradient β in the regression line $y = \alpha + \beta x$. We first calculate the corrected sums of squares and products; the results obtained appear on the following page.

$$S_{xy} = \Sigma xy - \frac{(\Sigma x)(\Sigma y)}{n} = 10\,596{\cdot}0 - \frac{504{\cdot}06 \times 182{\cdot}1}{9} = 397{\cdot}166,$$

$$S_{xx} = \Sigma x^2 - \frac{(\Sigma x)^2}{n} = 30\,162{\cdot}5 - \frac{(504{\cdot}06)^2}{9} = 1931{\cdot}79$$

and

$$S_{yy} = \Sigma y^2 - \frac{(\Sigma y)^2}{n} = 3830{\cdot}09 - \frac{(182{\cdot}1)^2}{9} = 145{\cdot}60.$$

The estimate of β is, therefore, $\hat{\beta} = 397{\cdot}166/1931{\cdot}79 = 0{\cdot}2056$. To determine whether the dependent variable y is related to the explanatory variable x we need to test the hypothesis H_0: $\beta = 0$, and to do this we need to specify the statistical model more precisely. Employing the model

$$E(Y_i) = \alpha + \beta x_i,\ i = 1, 2, \ldots, n,\ \mathrm{Var}(Y_i) = \sigma^2,\ i = 1, 2, \ldots, n,$$

with different Ys independently normally distributed, we then obtain

$$\hat{\beta} \sim N(\beta, \sigma^2/S_{xx}).$$

The unknown variance σ^2 is estimated conventionally by s^2, where

$$s^2 = \frac{1}{(n-2)}\left(S_{yy} - \frac{S_{xy}^2}{S_{xx}}\right),$$

and, by an ılogy with inference about the mean of a single random sample, we obtain

$$\frac{\hat{\beta} - \beta}{s/\sqrt{S_{xx}}} \sim t_{n-2}.$$

We thus test the hypothesis $\beta = 0$ by comparing this quantity, in which β is replaced by the null hypothesis value 0, with percentage points of the t-distribution. Here $\hat{\beta} = 0{\cdot}2056$, $s^2 = 9{\cdot}135$ and $S_{xx} = 1931{\cdot}79$, so that the statistic is 2·990. Comparing this with tables of the t-distribution on 7 degrees of freedom we find that the 5% and 1% critical values for a two-tailed test are 2·36 and 3·50 respectively. The result is thus significant at the 5% level, though not at the more stringent 1%, and, rejecting the hypothesis that $\beta = 0$, we conclude that the proportion of Arts students is relevant to the unemployment rate.

Notes

(1) The mathematical aspects of the solution above have been written out in some detail, but the practical problems of interpretation in problems like these are often subtle and easy to overlook. In the present case there are many questionable elements, including the following:

(i) normality; is it plausible?

(ii) linearity of relationship: how plausible is that?

(iii) is it reasonable to judge the relationship from records of 9 of the 18 universities without medical schools?

(iv) would there not be many more possible explanations for variations in unemployment rates which do not directly involve the proportion of Arts students?

* (2) The calculation of s above is a little cumbersome, and the presentation of an Analysis of Variance table often helps to formalise the process – and indeed illustrates the links regression has with one-way and two-way analysis of variance. (Section 5B deals with these topics.) For the present case the table is as follows.

Source	Degrees of Freedom	Sum of Squares	Mean Square
Regression	1	81·66	81·66
Residual	7	63·94	9·135
Total	8	145·60	

The estimate s^2 of σ^2 is then the Residual Mean Square, and the test of the hypothesis $\beta = 0$ can alternatively be performed by taking the ratio $81{\cdot}66/9{\cdot}135 = 8{\cdot}939$, and comparing it with tables of the F-distribution on 1 and 7 degrees of freedom. (Mathematically, the square of a random variable with Student's t-distribution has an F-distribution, as we see in Note 1 to Problem 2B.7. In this case the t-statistic is 2·9898, whose square is 8·939. We see that the tests in the solution and in this note are identical, and are not competitors.)

(3) The wording of the problem, with its use of 'depend', justifies the employment of a two-tailed test. Yet these matters are in many cases not all that clear cut. Since 'everybody knows' that Science graduates are more employable than Arts graduates (a point hinted at by the exclusion of universities with medical schools), it is at least arguable that one is not really interested in a negative regression coefficient, and that a one-tailed test could be justified. A sensible approach to a problem of this sort is to try to take a reasonable view of the intention of the problem, and then to justify the decision taken, one-tail or two-tail, in words. But it is important to remember that it is the intention of the problem, and not the appearance of the data, which matters. It is bad practice to decide which test to perform after looking at the data.

(4) While it is not required for this particular problem, it is worth noting that once $\hat{\beta}$ has been obtained one can straightforwardly calculate

$$\hat{\alpha} = \bar{y} - \hat{\beta}\bar{x} = 20{\cdot}23 - 56{\cdot}01 \times 0{\cdot}2056 = 8{\cdot}72,$$

so that the regression line is $y = 8{\cdot}72 + 0{\cdot}2056x$.

5A.3 Regression for several sets of data

The data below are claimed to relate to four experiments, which were conducted with the objective of investigating the effect of the value of a variable x on the associated value of a variable y. In each experiment the same values of x were used: thus only one column of x-values is given, while there are four columns of y-values, one for each experiment.

x	y_1	y_2	y_3	y_4
10·0	21·26	22·40	20·70	22·94
8·0	20·57	21·84	20·43	19·85
13·0	20·15	21·35	25·31	19·65
9·0	22·28	22·15	20·56	20·25
11·0	21·36	22·27	20·80	22·38
14·0	22·30	20·43	21·19	23·84
6·0	21·35	20·25	20·21	20·55
4·0	18·81	17·66	19·90	19·53
12·0	23·63	21·90	20·95	21·55
7·0	18·73	21·15	20·31	19·72
5·0	20·01	19·05	20·09	20·19

For each set of data:

(i) obtain the equation of the least squares regression line of y on x;

(ii) plot the data, and comment on the results.

Solution

(i) Our basic calculations for the first set of data are as follows:

$$\sum_{i=1}^{11} x_i = 99{\cdot}0, \qquad \sum_{i=1}^{11} y_i = 230{\cdot}45,$$

$$\sum_{i=1}^{11} x_i^2 = 1001{\cdot}0, \qquad \sum_{i=1}^{11} y_i^2 = 4850{\cdot}3015,$$

$$\sum_{i=1}^{11} x_i y_i = 2104{\cdot}85.$$

Corrected sums of squares and products are then

$$S_{xx} = 1001{\cdot}0 - \frac{99{\cdot}0^2}{11} = 110{\cdot}0,$$

$$S_{yy} = 4850{\cdot}3015 - \frac{230{\cdot}45^2}{11} = 22{\cdot}374,$$

$$S_{xy} = 2104{\cdot}85 - \frac{99{\cdot}0 \times 230{\cdot}45}{11} = 30{\cdot}8.$$

We continue to find estimates of the slope β and intercept α of the regression line as follows:

$$\hat{\beta} = \frac{30{\cdot}8}{110{\cdot}0} = 0{\cdot}28,$$

$$\hat{\alpha} = \frac{230{\cdot}45}{11} - 0{\cdot}28 \times \frac{99{\cdot}0}{11} = 18{\cdot}43.$$

When we turn to the other three data sets, we find that the results of the basic calculations, and therefore of the whole analysis, are the same.

(ii) Plots of the four data sets are presented in Figure 5.4.

The plot of the first data set reveals nothing untoward. The other three data sets, however, result in plots which indicate that the statistical analysis we have performed is inappropriate. The plot of the second data set indicates a strong curvilinear relationship: a possible analysis for this data set would be one in which the regression of y on x is quadratic (see Note 2 to Problem 5A.1). In a plot of the third data set all but one of the points lie close to a straight line, the other being far above the line. Such an observation (an *outlier*) merits our attention: in this case the best approach might be to reanalyse the data with this observation deleted (see Note 3). The appearance of the fourth data set suggests that the variability of y increases with the corresponding value of x. In such a case the method of least squares, which we have used to fit the regression line, is inappropriate, since it attaches equal weight to all observations in fitting the regression line.

Notes

(1) The fact that all four experiments yield results leading to exactly the same analysis is clearly suspicious. The data in this question have been artificially constructed to illustrate the point that an unthinking analysis of data can, through the application of inappropriate methods, lead to conclusions which are seriously in error. When contemplating any statistical analysis, it is important to check that the assumptions underlying that analysis are, at least approximately, valid for the data in hand. For bivariate data, where an analysis based on correlation or

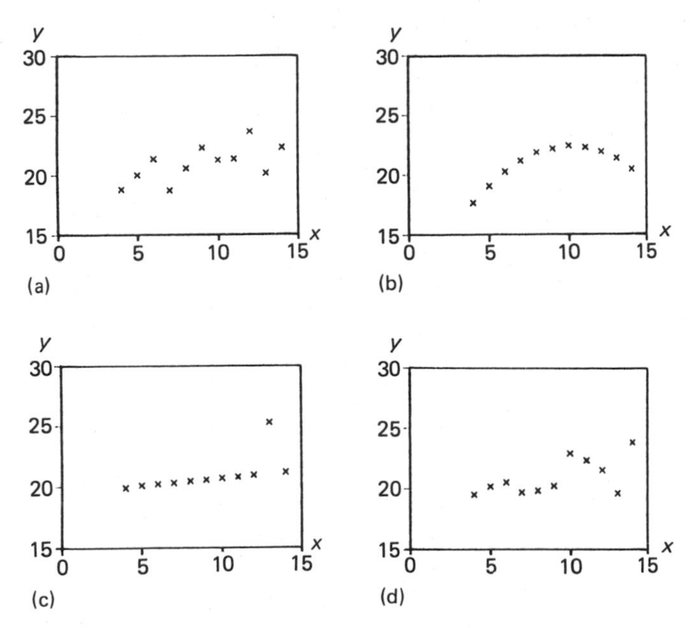

Figure 5.4 Plots for data of Problem 5A.3
(a) y_1 against x (b) y_2 against x (c) y_3 against x (d) y_4 against x

regression techniques is contemplated, a simple plot is an important preliminary to (or part of) the analysis. (We are referring here, of course, to 'real life', rather than to the more artificial process of answering examination questions, in which plots, unless specifically asked for, are usually something of a luxury.)

Three of the data sets in this problem are based on sets originally presented by F. J. Anscombe in an article published in 1973 in *The American Statistician*. These have been modified so that the basic calculations for regression yield precisely the same results for each data set, whereas results for the original data sets, while in close agreement, were not identical. It can be instructive to conduct a class exercise in which students are each given one of the four data sets and asked to fit a regression line. Comparing numerical results they should find that their answers agree exactly, but when they plot the data they find that they have been analysing different sets, and that some have been doing an inappropriate analysis. Thus they learn, through experience, the value of plotting data before any further analysis.

(2) The problem could have asked for a more detailed analysis, involving questions such as those posed in Problems 5A.1 and 5A.2, but, in view of the artificial nature of the data, we did not feel this was justified. It is worth remarking, however, that if we had asked for any further analysis along these lines the results would have been the same for all four data sets. The same would apply if we were to set a problem involving the production of (product moment) correlation coefficients for the four data sets.

(3) The problem of outliers is encountered in all areas of statistics, and not just in regression. It is not good practice for a statistician to discard observations just because they 'don't look right'. But when outlying observations are present, including them in a standard analysis can render that analysis meaningless, as can easily be seen by superimposing the regression line $y = 18{\cdot}43 + 0{\cdot}28x$ on a scatterplot of y_3 against x. We frequently resolve this difficulty not by *discarding* any outliers, but by reporting separately on them and then analysing and reporting on the remaining data in the usual way. Before this is done, however, each offending observation should be scrutinised to see if, for example, there has been a simple error in transcribing a figure (in which case the error can be corrected and the data reanalysed), or to

see if the experimenter has any explanation for the discrepancy (in which case the observation might be omitted from further formal analysis). Sometimes such scrutiny reveals nothing, in which case there may be no alternative to deleting (and reporting separately on) the offending observation(s). If this course of action has to be considered, informal judgments as to whether suspicious-looking observations are outliers or not may be supplanted by formal statistical tests designed for this purpose. In the case of the third data set in this problem the picture is sufficiently clear for no formal test to be necessary.

(4) A fifth data set which gives precisely the same results as the four in this problem has ten y-values, 21·20, 18·74, 21·98, 19·70, 22·66, 19·32, 21·62, 20·91, 20·13 and 20·44, all corresponding to $x = 8$, and an eleventh y-value of 23·75 corresponding to $x = 19$. Though this data set is less well suited to a class exercise of the type described in Note 1, it illustrates another important statistical point. The observation corresponding to $x = 19$ is *influential*: whereas even a quite large change in any of the other y-values would have only a slight effect on the fitted regression line, a change in this value would noticeably alter the fitted line. Thus the accuracy of our estimates depends crucially on the reliability of a single observation; this would generally be regarded as an unsatisfactory state of affairs. (Note that there is no suggestion that a standard regression analysis is inappropriate for this data set. Our objection here is to the *design* of the experiment giving rise to the data; indeed, a further objection to the design is that it does not permit us to detect situations, such as those illustrated by the second and fourth data sets, in which a standard analysis would be inappropriate.)

5A.4 Inflation, money supply and dysentery

The table below gives the values of the following quantities over the years 1967-1975.

x: The annual inflation rate in the U.K. (measured as a percentage).

y: The excess money supply two years previously (measured as a percentage).

z: The number (in thousands) of cases of dysentery reported in Scotland during the previous year.

	x	y	z
1967	2·5	4·7	4·3
1968	4·7	1·9	4·5
1969	5·4	7·8	3·7
1970	6·4	4·0	5·3
1971	9·4	1·3	3·0
1972	7·1	7·8	4·1
1973	9·2	11·4	3·2
1974	16·1	23·4	1·6
1975	24·2	22·2	1·5

(a) Calculate the product moment correlation coefficient between variables x and y and demonstrate that the value you have obtained differs significantly from zero.

(b) Do the same for variables x and z.

(c) These data were presented in letters which appeared in the correspondence columns of *The Times* in April, 1977. The writer of one letter suggested that the values of x and y demonstrated clearly the effect of the money supply on inflation, while the writers of a later letter maintained that those of x and z showed clearly the effectiveness of Scottish dysentery in keeping prices down. Comment on these claims.

Solution

(a) The basic calculations $\Sigma x = 85{\cdot}0$, $\Sigma y = 84{\cdot}5$, $\Sigma x^2 = 1166{\cdot}72$, $\Sigma y^2 = 1335{\cdot}43$ and $\Sigma xy = 1174{\cdot}86$ lead to corrected sums of squares and products $S_{xx} = 363{\cdot}94$, $S_{yy} = 542{\cdot}07$ and $S_{xy} = 376{\cdot}80$. The correlation coefficient between x and y is therefore

$$r_{xy} = \frac{376{\cdot}80}{\sqrt{363{\cdot}94 \times 542{\cdot}07}} = 0{\cdot}848.$$

To demonstrate that this value differs significantly from zero we note, if appropriate tables are available, that the critical value of the correlation coefficient based on a sample of size 9 for a two-sided test at the 1% significance level is $0{\cdot}7977$. Since our observed value of r_{xy} exceeds this critical value it may be judged to be highly significant.

If tables of critical values for the correlation coefficient are not available, we calculate

$$t = 0{\cdot}848\sqrt{\frac{7}{1{\cdot}0 - 0{\cdot}848^2}} = 4{\cdot}239.$$

Comparing this value with critical values of the t-distribution with 7 degrees of freedom, we find that it exceeds the (two-tailed) 1% point ($3{\cdot}4995$); again, we judge our observed correlation to be highly significant.

(b) Our basic calculations follow the same pattern as that followed in the solution to part (a); some quantities, however, have been calculated already and details will not be repeated here. New quantities required are $\Sigma z = 31{\cdot}2$, $\Sigma z^2 = 121{\cdot}38$ and $\Sigma xz = 234{\cdot}61$, leading (along with results obtained earlier) to $S_{zz} = 13{\cdot}22$ and $S_{xz} = -60{\cdot}06$. Recalling that $S_{xx} = 363{\cdot}94$, we find that the correlation coefficient between x and z is

$$r_{xz} = \frac{-60{\cdot}06}{\sqrt{363{\cdot}94 \times 13{\cdot}22}} = -0{\cdot}866.$$

As before we note, if tables of critical values of the correlation coefficient are available, that our observed value of r_{xz}, being less than $-0{\cdot}7977$, is significant at the 1% level. Alternatively, we calculate

$$t = -0{\cdot}866\sqrt{\frac{7}{1{\cdot}0 - 0{\cdot}866^2}} = -4{\cdot}578.$$

Once again, since $|t| > 3{\cdot}4995$, we reject the hypothesis of zero correlation in the underlying population, testing at the 1% significance level.

(c) In linking the rate of inflation to the reported prevalence of dysentery in Scotland, the authors of the second letter, while not necessarily disagreeing with the suggestion of a link between the money supply and the rate of inflation, were making the point that the conclusion of the earlier letter should be treated with extreme caution. To draw attention to the fact that the misuse of statistical techniques (and correlation in particular) can lead to 'proofs' of erroneous hypotheses, they produced a statistical argument leading to a conclusion which was clearly ridiculous.

The fundamental error in arguments, based on data such as these, leading to the conclusion that 'y causes inflation' (where 'y' represents any possible cause, not necessarily the particular quantity discussed above), is that the demonstration of *association* between y and inflation does not imply that y in fact *causes* inflation. We are dealing with observational, rather than experimental, data and it is possible (and, indeed, likely) that both y and the rate of inflation are affected by other factors.

Notes

(1) In correlation studies we treat the two variables symmetrically, and all that we can demonstrate is an association. Here we are tempted to interpret the association in terms such as 'y causes inflation', rather than 'inflation causes y', merely because the the observations of y precede the corresponding observations of the inflation rate.

(2) Examples of 'spurious correlation' like the above, where the observations are gathered through time, frequently arise because there is an underlying time trend in the variation of the two quantities under observation. That this is the case in this example is easily seen by, for example, plotting the variables against time. Time, indeed, provides a better 'explanation' of variations in the inflation rate than either of the other variables: the correlation coefficient between the rate of inflation and the year is 0·875.

The presence of a time trend means that correlation techniques are inappropriate for the data of this problem, since we do not have a random sample from a bivariate distribution. Unfortunately, public examination questions which ask the candidate to make inappropriate use of correlation coefficients are quite common, and this fact justifies our inclusion of the problem in this book. (Calculating a correlation coefficient between the rate of inflation and time at the end of the preceding paragraph is also, strictly speaking, invalid – see Note 1 to Problem 5A.1. We are, however, using the correlation coefficient here merely as a numerical measure on the basis of which to compare apparent strength of relationships.)

(3) The two tests for significance of the correlation coefficient used in our solution are equivalent. A partial demonstration of their equivalence is possible if we consider, for example, a value of r equal to the 1% critical value, 0·7977. This transforms to a t-value of

$$0{\cdot}7977\sqrt{\frac{7}{1{\cdot}0 - 0{\cdot}7977^2}} = 3{\cdot}4997$$

which, to the accuracy we are justified in claiming given that the value we are transforming is accurate to four significant figures, agrees with the critical value obtained from t-tables.

A test based on the z-transformation of the correlation coefficient is also possible: however, since it is based on an approximation, the exact tests that we have used are preferred. See Note 2 to Problem 5A.5 for a comparison between the exact and approximate tests.

(4) We have seen in Note 2 that correlation techniques are inappropriate here. Regression techniques (with the rate of inflation as the response variable) would be more appropriate, since they do not require a random sample from a bivariate distribution, and the aim of regression is to *explain* variation in one quantity in terms of another. The same problems arise, however, since we are dealing with observational data.

The close links between the techniques of regression and correlation may be illustrated if we consider the application of regression techniques to the data of this problem. If, in particular, we fit a regression line with y as the dependent variable and x as the explanatory variable, we obtain (as in Problems 5A.1 and 5A.2) $\hat{\beta} = 1{\cdot}035$ and $s^2 = 21{\cdot}71$. A test of the null hypothesis $\beta = 0$ may then be based on a t-statistic with the value $\dfrac{1{\cdot}035}{\sqrt{21{\cdot}71/363{\cdot}94}} = 4{\cdot}239$ which, we note, is the value obtained in the solution to part (a).

5A.5 Correlation between height and weight

The heights (in cm) and weights (in kg) of thirteen male and ten female students are given below:

Male				Female			
Height	Weight	Height	Weight	Height	Weight	Height	Weight
183	65·5	190	79·5	170	72·5	160	54·0
170	64·0	182	62·0	165	51·0	160	65·5
178	64·0	192	86·0	168	57·5	157	56·5
173	66·0	173	76·5	163	44·5		
175	58·5	175	64·0	160	52·0		
178	67·0	188	64·0	155	59·5		
185	72·0			168	64·5		

(a) Obtain 95% confidence intervals for the correlation coefficients between height and weight for the male and female students.

(b) Is there any evidence in the data to suggest that the two correlation coefficients differ?

Solution

(a) We shall let h denote height, and w weight. Starting with the male students, standard calculations lead to $\Sigma h = 2342{\cdot}0$, $\Sigma w = 889{\cdot}0$, $\Sigma h^2 = 422\,522{\cdot}0$, $\Sigma w^2 = 61\,538{\cdot}0$ and $\Sigma hw = 160\,527{\cdot}5$, and hence to $S_{hh} = 601{\cdot}6923$, $S_{ww} = 744{\cdot}0769$ and $S_{hw} = 370{\cdot}7308$. The sample correlation coefficient between height and weight for the male students is therefore

$$r_m = \frac{370{\cdot}7308}{\sqrt{601{\cdot}6923 \times 744{\cdot}0769}} = 0{\cdot}554.$$

To obtain a confidence interval for the correlation coefficient ρ_m between height and weight for male students, we use Fisher's z-transformation. The transformed value of r_m is

$$z_m = \frac{1}{2}\log_e\left(\frac{1{\cdot}0 + 0{\cdot}554}{1{\cdot}0 - 0{\cdot}554}\right) = 0{\cdot}624.$$

The distribution of z_m is approximately normal with variance $1/(n-3)$, where n is the sample size. A 95% confidence interval for the transformed value of ρ_m therefore has end-points

$$0{\cdot}624 \pm \frac{1{\cdot}96}{\sqrt{10}},$$

i.e. $0{\cdot}004$ and $1{\cdot}244$. Applying the inverse transformation to the upper limit we find, as the upper limit of our confidence interval for ρ_m,

$$\frac{e^{2\times1{\cdot}244} - 1{\cdot}0}{e^{2\times1{\cdot}244} + 1{\cdot}0} = 0{\cdot}847.$$

Applying the same transformation to the lower limit, we find that the transformed value is the same as the untransformed value (since the z-transformation has very little effect on values close to zero). Our 95% confidence interval for ρ_m is therefore $(0{\cdot}004, 0{\cdot}847)$.

Turning now to the female students, we obtain $\Sigma h = 1626{\cdot}0$, $\Sigma w = 577{\cdot}5$, $\Sigma h^2 = 264\,616{\cdot}0$, $\Sigma w^2 = 33\,946{\cdot}75$ and $\Sigma hw = 94\,022{\cdot}5$, leading to $S_{hh} = 228{\cdot}40$, $S_{ww} = 596{\cdot}125$ and $S_{hw} = 121{\cdot}00$. The sample correlation coefficient between height and weight for the female students is therefore

$$r_f = \frac{121{\cdot}00}{\sqrt{228{\cdot}40 \times 596{\cdot}125}} = 0{\cdot}328.$$

To obtain a confidence interval for the correlation coefficient ρ_f between height and weight for female students, we proceed as we did for male students. We obtain

$$z_f = \frac{1}{2}\log_e\left(\frac{1{\cdot}0 + 0{\cdot}328}{1{\cdot}0 - 0{\cdot}328}\right) = 0{\cdot}340.$$

The 95% confidence interval for the transformed correlation coefficient has end-points

$$0{\cdot}340 \pm \frac{1{\cdot}96}{\sqrt{7}},$$

i.e. $(-0{\cdot}400, 1{\cdot}081)$. The corresponding interval for ρ_f is $(-0{\cdot}380, 0{\cdot}794)$.

(b) A test of the equality of ρ_m and ρ_f may be based on the difference between the observed values of z_m and z_f: if the two population correlation coefficients are the same, the distribution of $z_m - z_f$ is approximately $N(0, 1/10 + 1/7)$. A test of the null hypothesis that the two population coefficients are equal may therefore be performed by comparing the statistic

$$\frac{z_m - z_f}{\sqrt{1/10 + 1/7}}$$

with critical values of the standard normal distribution; since the observed value of this statistic is 0·576, there is no reason to reject the null hypothesis.

Notes

(1) Two points about the z-transformation are worth mentioning. The first is the obvious one that, despite the notation which is in almost universal use, the distribution that is used to approximate that of z is not the *standard* normal distribution $N(0,1)$. The second is that, although we have indicated in the solution to this problem how the z-transformation and its inverse can be computed on a calculator with exponential and logarithmic functions, some readers may prefer to use tables of the two transformations. It is the common practice for such tables to deal only with positive values of r and z, since negative values may be dealt with using the easily verifiable properties $z(-r) = -z(r)$ and $r(-z) = -r(z)$.

(2) The fact that the confidence interval obtained for ρ_m does not include the value zero indicates that, testing at the 5% significance level, the null hypothesis $\rho_m = 0$ should be rejected – but only just, since the lower limit of the interval lies quite close to zero. This suggests that, since our method based on the z-transformation is only approximate, it might be interesting to investigate the accuracy of the approximation by comparing with the results of an exact test. Doing so, we reach the same conclusion, since the appropriate critical value for the correlation coefficient obtained from tables is 0·5529. This suggests that, at least when the underlying correlation coefficient is close to zero, methods involving the z-transformation may provide a fairly good approximation to the exact distribution, even for quite small sample sizes. We can investigate this point further by finding the value of r which would, at the 5% significance level, have been judged 'just significant' by an approximate test based on the z-transformation. This would correspond to a value of $|z|$ of $1{\cdot}9600/\sqrt{10} = 0{\cdot}6198$, which transforms to a value of $|r|$ of 0·5510, quite close to the exact critical value quoted above.

(3) Since the (product moment) correlation coefficient for male students exceeds the 5% critical value by only a narrow margin, this example permits an interesting comparison with rank correlation methods. If, for example, we calculate Spearman's coefficient, we obtain the value 0·386 (adjusting for ties), or 0·397 (without adjustment). Since the 5% critical value for Spearman's coefficient, based on a sample of size 13, is 0·560, the observed value is not significant. The difference between the results of these two tests illustrates the loss of information resulting from the replacement of the actual values of height and weight by ranks.

5A.6 Judging a beauty contest

The twelve contestants in a beauty contest are judged by two judges, each of whom ranks the contestants in order. The results are as follows:

Contestant	1	2	3	4	5	6	7	8	9	10	11	12
Judge A	8	2	1	12	9=	5=	5=	9=	3	9=	7	4
Judge B	8=	2=	1	10	6	2=	12	8=	4	5	11	7

Calculate a rank correlation coefficient between the two judges' results and perform an appropriate significance test. Comment on your results.

Solution

Since the statement of the problem does not specify which rank correlation coefficient is to be used, we present two solutions, one involving Spearman's coefficient and one involving Kendall's.

Spearman's coefficient

Letting a and b respectively represent the ranks assigned by judges A and B, Spearman's coefficient can be calculated as the product moment correlation coefficient between the as and bs. (The tied ranks 2=, 5=, 8= and 9= are replaced by the values 2·5, 5·5, 8·5 and 10·0.) Performing the usual calculations, we obtain

$$S_{aa} = 140{\cdot}5, \; S_{bb} = 142{\cdot}0, \; S_{ab} = 78{\cdot}75.$$

Spearman's rank correlation coefficient is therefore

$$r_S = \frac{78{\cdot}75}{\sqrt{140{\cdot}5 \times 142{\cdot}0}} = 0{\cdot}558.$$

To assess the significance of this value, we consult tables for Spearman's coefficient, and find that the 5% and 10% critical values for a sample of size 12 are 0·5874 and 0·5035 respectively. (Although we are performing a two-tailed test, there is a fairly strong argument in favour of a one-tailed test – see Note 1.) Our observed value, therefore, differs from zero at the 10%, but not at the 5%, level of significance. There is thus a fairly strong indication that the two judges tend to agree (as contestants would hope), but the evidence is hardly conclusive: the calculated correlation coefficient reflects the fact (obvious on inspecting the data) that their agreement is far from total.

An alternative method of calculating r_S is to make use of the formula

$$r_S = 1 - \frac{6\sum_{i=1}^{n} d_i^2}{n(n^2-1)},$$

where n denotes the number of contestants and, for $i = 1, 2, \ldots, n$, d_i denotes the difference between the ranks a_i and b_i assigned to the ith contestant. This formula is exact if there are no ties in either ranking, but is only approximate otherwise; the approximation is, however, fairly good unless there are many ties. For our data, we obtain the values below.

a_i	8·0	2·0	1·0	12·0	10·0	5·5	5·5	10·0	3·0	10·0	7·0	4·0
b_i	8·5	2·5	1·0	10·0	6·0	2·5	12·0	8·5	4·0	5·0	11·0	7·0
d_i	−0·5	−0·5	0·0	2·0	4·0	3·0	−6·5	1·5	−1·0	5·0	−4·0	−3·0

We now find that $\sum_{i=1}^{12} d_i^2 = 125{\cdot}0$, so that

$$r_S \approx 1{\cdot}0 - \frac{6 \times 125{\cdot}0}{12 \times 143} = 0{\cdot}563.$$

As can be seen, the approximation performs well for this example.

Kendall's coefficient

To calculate this coefficient, we consider all $n(n-1)/2$ pairs of contestants. For each pair we score +1 if the two judges rank the two contestants in the same way, and −1 if they are ranked differently: if the two contestants are awarded the same rank by either or both of the judges the score is 0. If the total of the scores obtained is S, Kendall's coefficient r_K is defined as the ratio

$$r_K = \frac{S}{\sqrt{N_A N_B}},$$

where N_A and N_B are the numbers of pairs ranked distinctly by A and B respectively. If there were no ties in the judges' rankings both N_A and N_B would be equal to $n(n-1)/2$, i.e. to 66 in this problem. As it is, the members of four pairs are tied by A (one pair having the rank 5·5, while three pairs have the rank 10), and there are two pairs tied by B. Thus $N_A = 62$, and $N_B = 64$.

One way of calculating S involves first rearranging the observations from left to right so that the ranks awarded by one of the judges (it does not matter which) are in nondecreasing order. We then work through the contestants from left to right, considering the ranks awarded by the other judge. For each contestant we consider the ranks appearing to the right of his or her position in the list, excluding those corresponding to contestants tied with the current contestant by the first judge. We score +1 for higher ranks, and −1 for lower ranks; noting the contribution to the total score, we move on to the next contestant. If we choose Judge A as the first judge, results are as follows.

Judge A	1	2	3	4	5·5	5·5	7	8	10	10	10	12
Judge B	1	2·5	4	7	2·5	12	11	8·5	8·5	6	5	10
Score	11	9	7	2	6	−6	−5	−1	1	1	1	0

(It may be in order to explain how some of the above scores are obtained. The first is obtained by considering all ranks to the right of the '1' in the 'Judge B' row. Since all eleven of these exceed 1, the score of 11 is obtained. The eighth score is obtained by considering the four ranks appearing to the right of the first '8·5' in that row. Of these, one exceeds 8·5, and two are less than this value, so that the score is $1-2=-1$. The ninth is obtained by considering the three ranks appearing to the right of the '8·5' in the row of Judge B's ranks, but excluding the ranks 6 and 5 due to ties in Judge A's ranking: this leaves just the rank 10, leading to a score of $1-0=1$.)

Totalling the scores obtained, we find that $S = 26$; thus

$$r_K = \frac{26}{\sqrt{62 \times 64}} = 0{\cdot}413.$$

Comparing the value of r_K with tables, we find that it differs significantly from zero at the 10% level, but not at the 5% level (the appropriate critical values are 0·3939 and 0·4545 respectively). Our conclusions are thus similar to those reached through the calculation of r_S.

As with Spearman's coefficient, there is an alternative method of calculation for r_K which is only approximate in the presence of ties, but which is often used unless there are many ties. The (slight) simplification in this method is that we ignore ties in calculating N_A and N_B, so that both have the value $n(n-1)/2$. With this simplification, for our data, we have

$$r_K \approx \frac{26}{66} = 0{\cdot}394.$$

We note that the effect of ignoring ties, though greater than in the case of Spearman's coefficient, is slight.

Notes

(1) Although in most hypothesis testing problems it is fairly clear whether a one- or two-tailed test is appropriate (and a fairly good rule is to use the latter unless the wording of the problem indicates that departures from the null hypothesis in a particular direction are of interest), there are some problems where the situation is rather less clear-cut. In such situations our view would be that either type of test would be valid, *provided* that some justification is given for the choice. In this problem, despite the absence of any wording which suggests a one-tailed test, there is a strong argument in favour of a test which rejects the null hypothesis for large positive values of the correlation coefficient, since such values indicate some agreement between the two

judges and this is what we would hope to detect. Large negative values indicate systematic disagreement, which could lead us to suspect that one or other of the judges is worse than useless!

Despite this argument, we have chosen to perform a two-tailed test, on the grounds that, if there is systematic disagreement, we would be interested in detecting and, if possible, explaining it.

(Note that there is nothing in the statement of the problem to tell us whether a (numerically) high rank corresponds to a good or a bad candidate. It is, of course, not necessary to know this *provided* that the two judges are following the same convention in ranking the candidates. If they did not, we would expect to observe results suggesting a degree of systematic disagreement.)

(2) The presence of ties in the rankings causes two problems. The first is that computational formulae which do not correct for ties lead to results which are slightly inaccurate. This can, of course, be overcome (as in our solution) by using the correct formulae: this does not add much to the work involved in the case of Kendall's coefficient, and need not add much in the case of Spearman's coefficient if a suitable calculator is available.

The second problem resulting from the presence of ties in the rankings concerns the tables of critical values for the two coefficients. These are based on the distributions which arise if the ranks awarded by each judge are independent random permutations of the integers $1, 2, \ldots, n$. The tables are thus, strictly speaking, inappropriate in the presence of ties. There is no totally satisfactory way round this problem, and all we can do is hope that the presence of relatively few ties has only a slight effect on the accuracy of the tabulated values.

(3) If tables for testing the significance of rank correlation coefficients are not available, approximations may be used. For Spearman's coefficient it is possible to use tables for the product moment correlation coefficient (or, equivalently, to transform the value of r_S to a value which may be checked against t-tables, as in Problem 5A.4). The approximation is fairly good even for quite small values of n: for $n = 12$ we can compare the exact 5% critical value used in our solution (0·5874) with the corresponding value for the product moment correlation coefficient (0·5760).

For Kendall's coefficient a normal approximation can be used: in the absence of association, the distribution of r_K is approximately normal with mean zero and variance $2(2n+5)/9n(n-1)$. For a good approximation, however, we require a value of n rather larger than we have here: the exact 10% critical value (0·3939) may be compared with the approximation $1{\cdot}6449\sqrt{58/1188} = 0{\cdot}3635$.

(4) A concise expression for the quantity S which appears in the definition of Kendall's coefficient is

$$S = \sum_{\substack{i=1 \\ i<j}}^{n} \sum_{j=1}^{n} \operatorname{sign}(a_i - a_j)\operatorname{sign}(b_i - b_j),$$

where

$$\operatorname{sign}(x) = \begin{cases} 1 & \text{if } x > 0 \\ 0 & \text{if } x = 0 \\ -1 & \text{if } x < 0. \end{cases}$$

It is possible to derive a similar expression for S_{ab}, viz.

$$S_{ab} = \frac{1}{n} \sum_{\substack{i=1 \\ i<j}}^{n} \sum_{j=1}^{n} (a_i - a_j)(b_i - b_j).$$

This shows that Kendall's coefficient is similar to Spearman's coefficient, but makes rather less use of the information provided by the ranks, just as Spearman's coefficient is a version of the product moment correlation coefficient which replaces actual data values by ranks.

5B Analysis of Variance

This topic is mainly covered at undergraduate level, where it can play a very prominent rôle. It does have a very wide application, especially in industrial and agricultural research, and is therefore of particular value through the insights it offers into the practical applications of statistics.

The method was devised by Sir Ronald A. Fisher, and is just one of his fundamental contributions to statistics. Its main application is to what are called *comparative experiments*; these are experiments in which we may compare, for example, the yields of two or more types of crop, or the fuel consumption of different models of car, or measures of hand-eye coordination of children of different ages. In each of these cases, we obtain experimental data. Roughly speaking, the analysis always proceeds by measuring the total variability in the data (by the corrected sum of squares of all the data values), and then allocating this to the various possible sources of that variability. In this way one can judge the importance or otherwise of the various sources, and draw appropriate conclusions.

At its simplest, the analysis of variance can be viewed as little more than an extension of the two-sample t-test to more than two samples. But the topic is much richer than that. Two-way analysis of variance (and, indeed, generalisations which we do not cover here) shows the power of statistical methods of conducting and analysing experiments so as to draw conclusions simultaneously about two types of variation. Thus, in the crop yield example above, one would not in practice conduct an experiment solely to compare yields of a few types of crop. In all commercial farming, fertilisers are applied to crops, and experiments would be performed to obtain information simultaneously about crops and fertilisers. The idea of experimenting in this way was considered rather unsound when Fisher first put it forward, but is now recognised as being one of the most important contributions of statistics to scientific experimentation.

5B.1 Relationship between quality and temperature

An industrial plant was maintained at different temperatures on four successive days, and on each day three samples were taken from the process and analysed for quality; a score was awarded in arbitrary units, and the resulting scores are given in the table below.

Sample	Temperature (Day)			
	$100°C$ (1)	$120°C$ (2)	$140°C$ (3)	$160°C$ (4)
1	41	54	50	38
2	44	56	52	36
3	48	53	48	41

Is there evidence to suggest that average score depends on temperature? If so, determine between which temperatures there are significant differences at the 5% level.

Solution

Although the data appear in the form of a two-way table, the three scores in each column are, in fact, simply repeated (and presumed independent) samples at the same temperature. Hence a one-way analysis of variance model is appropriate, and we aim to partition the total variability of the data, as measured by the corrected sum of squares about the sample mean, into two components, a 'between temperatures' component and a 'within temperatures' component. This is done in the Analysis of Variance table which appears on the opposite page; the necessary calculations are given separately at the end of the solution.

Source	Degrees of Freedom	Sum of Squares	Mean Square	F-ratio
Between temperatures	3	434·25	144·75	23·16
Within temperatures	8	50·00	6·25	
Total	11	484·25		

The calculated F-ratio, 23·16, is now compared with the tabulated upper percentage points of the F-distribution on 3 and 8 degrees of freedom. At the 1% significance level the tabulated value is 7·59. The result is thus highly significant, giving very strong evidence that temperature affects score.

To discover more about the precise way in which score is affected by temperature, we can calculate the so-called Least Significant Difference (LSD) at the 5% level. We obtain

$$\text{LSD} = 2{\cdot}306\sqrt{2\times6{\cdot}25/3} = 4{\cdot}71;$$

in this formula 2·306 is the upper $2\frac{1}{2}$% point of the t-distribution on 8 degrees of freedom, and 6·25 is the mean square within temperatures.

The sample means for the four temperatures are now compared to see which pairs differ by more than the LSD. The means are (in ascending order of temperature) 44·33, 54·33, 50·00 and 38·33. We see then that the means for temperatures 120° and 140° are not significantly different at the 5% level, but all other pairs are significantly different at that level, and in particular the means for 100° and 160° are lower than the other two. We conclude that the mean score is highest for the two intermediate temperatures.

Calculations

For problems in the Analysis of Variance the calculations are rather more lengthy than in most others in this book, and it is convenient to set out the working leading to the Analysis of Variance table here. A satisfactory sequence of calculations is presented below.

(i) Grand total, $GT = (41 + 44 + \ldots + 36 + 41) = 561$; number of observations, $n = 12$.

(ii) Correction factor, $CF = (GT)^2/n = 561^2/12 = 26\,226{\cdot}75$.

(iii) Totals for temperatures: $T_1 = 133,\ T_2 = 163,\ T_3 = 150,\ T_4 = 115$.

Numbers of observations: $n_1 = n_2 = n_3 = n_4 = 3$.

(iv) Uncorrected total sum of squares, $USS = 41^2 + 44^2 + \ldots + 36^2 + 41^2 = 26\,711$.

(v) (Corrected) total sum of squares, $CSS = USS - CF = 26\,711 - 26\,226{\cdot}75 = 484{\cdot}25$.

(vi) Between temperatures sum of squares, $SSB = \frac{T_1^2}{n_1} + \frac{T_2^2}{n_2} + \frac{T_3^2}{n_3} + \frac{T_4^2}{n_4} - CF$

$$= \frac{1}{3}(133^2 + 163^2 + 150^2 + 115^2) - CF = 26\,661 - 26\,226{\cdot}75 = 434{\cdot}25.$$

(vii) Within temperatures sum of squares $= CSS - SSB = 484{\cdot}25 - 434{\cdot}25 = 50{\cdot}00$.

(viii) Degrees of freedom:

Total $= n - 1 = 11$,

Between temperatures $= 4 - 1 = 3$,

Within temperatures $= 11 - 3 = 8$.

(ix) Mean square = sum of squares / degrees of freedom.

(x) F-ratio $= 144{\cdot}75/6{\cdot}25 = 23{\cdot}16$.

Notes

(1) As in most statistical analyses, the data are assumed to satisfy some theoretical model and calculations are done to determine whether the model is correct. In this problem we assume that the three scores observed at each temperature are a random sample from a normal distribution with a particular mean and variance. We may denote the mean of the distribution of scores at temperature 100° by μ_1, the mean at temperature 120° by μ_2, etc. Then, if all temperatures produce different mean scores, we have made 3 independent observations on each of the distributions $N(\mu_1, \sigma^2)$, $N(\mu_2, \sigma^2)$, $N(\mu_3, \sigma^2)$ and $N(\mu_4, \sigma^2)$. Notice that the variance, σ^2, is assumed to be the same in all four distributions – this is a very important assumption, and would, of course, be checked in practice. The null hypothesis that we adopt is that $\mu_1 = \mu_2 = \mu_3 = \mu_4$ and we wish to test if this is true rather than the alternative hypothesis that at least two of these means are different. If the null hypothesis is true then the calculated F-ratio ought to have an F-distribution.

(2) The formula for the Least Significant Difference at level α is given by

$$\text{LSD} = t_{\nu,\alpha/2}\sqrt{2s^2/N} ,$$

where s^2 is the within temperatures mean square, and ν is its number of degrees of freedom, $t_{\nu,\alpha/2}$ is the appropriate t-distribution percentage point and N is the number of data values used to compute each of the means being compared.

The method derives from the conventional two-sample t-test for the difference between two means; a short discussion is given in Note 4 to Problem 4B.2.

(3) The LSD is a convenient and equivalent way of performing the six possible t-tests that would be required to compare each pair of means. The LSD method, however, ignores the fact that six hypothesis tests are being performed, not one, and so the significance level of 5% is an underestimate of the true chance of rejecting the null hypothesis when it is true. More advanced statistical techniques are available, however, for overcoming this difficulty.

(4) Although the LSD method is a legitimate one to use here, a better analysis can be obtained by making use of the fact that the four temperatures can be placed in a natural order, from low to high. Presumably the aim of the experiment was to discover which temperature is likely to maximise the mean score, and consequently there is a definite and intentional structure amongst the four temperatures. This structure has not been exploited in the given solution.

The underlying assumption that might have been made when planning the experiment was that score is likely to exhibit a roughly quadratic relationship with temperature: that is, if mean score were denoted by y and temperature by t then the relationship would be $y = a + bt + ct^2$. If a, b and c could be estimated, the maximum of this function, if it existed, could be obtained using calculus. With four equally spaced temperatures, as here, there are simple formulae for estimating a, b and c.

(5) In an experiment of the type described it is important that the conditions under which the industrial process is run are identical (except for temperature change) on each of the different days. Otherwise it is impossible to decide whether the differences between the four means are due to the temperature change or due to a change in the day. If the choice of day was thought to affect the score then a more involved experiment would be required.

5B.2 The rubber content of guayule plants

A random sample of 50 plants was selected from a field containing one-year-old plants of a variety of guayule, a plant species yielding rubber. Of the 50, 25 were classified as Normal (N), 14 were classified as Off-type (O) and 11 as Aberrant (A). The data on the following page show the percentage rubber content from each plant.

Normal				Off-type		Aberrant	
6·93	6·38	6·68	7·10	5·70	4·84	6·34	6·40
6·42	7·26	7·29	6·30	5·01	6·20	4·25	8·90
7·01	6·84	7·12	6·83	5·82	4·46	7·70	5·90
7·31	7·25	6·86	6·68	5·24	5·58	6·36	7·10
6·81	7·32	6·48		5·88	5·24	5·51	
7·00	6·42	6·20		6·04	5·58	4·72	
7·42	7·27	6·43		6·10	6·01	7·90	

(a) Test the null hypothesis that there is no difference in mean rubber content between the three types of plant.

(b) If the means of the distributions of each type of plant are denoted by μ_N, μ_O and μ_A respectively, test the following particular null hypotheses:

(i) $\mu_N = \mu_A$;

(ii) $\mu_O = \frac{1}{2}(\mu_N + \mu_A)$.

Solution

The data are classified in one way only and so the appropriate statistical model is that for a one-way analysis of variance. The Analysis of Variance table follows, the necessary calculations being laid out at the end of the solution.

Source	Degrees of Freedom	Sum of Squares	Mean Square	F-ratio
Between types	2	15·5457	7·7728	14·04
Within types	47	26·0182	0·5536	
Total	49	41·5639		

(a) The F-ratio for testing the null hypothesis that $\mu_N = \mu_O = \mu_A$ is 14·04 on 2 and 47 degrees of freedom. The tabulated upper 1% point of the F-distribution on these degrees of freedom is, by interpolation, 5·11. (Statistical tables are often less detailed for degrees of freedom higher than 40.) The F-ratio is highly significant, giving overwhelming evidence of a difference between the three types of plant.

(b) (i) To test the null hypothesis that $\mu_N = \mu_A$ we express the hypothesis as $\mu_N - \mu_A = 0$ and use a t-test. If $\bar{x}_N$ and $\bar{x}_A$ are the observed means of the two types and s^2 is the within types mean square, the t-statistic is

$$t = \frac{\bar{x}_N - \bar{x}_A}{\sqrt{s^2\left(\frac{1}{n_N} + \frac{1}{n_A}\right)}},$$

and has the t-distribution on 47 degrees of freedom. For the data in this problem, t is given by

$$t = \frac{6{\cdot}864 - 6{\cdot}462}{\sqrt{0{\cdot}5536\left(\frac{1}{25} + \frac{1}{11}\right)}} = 1{\cdot}50.$$

The $2\frac{1}{2}$% point of the t-distribution on 47 degrees of freedom is, again by interpolation, 2·01. On a two-tailed test at the 5% level, we therefore find no evidence to reject the null hypothesis.

(ii) To test the hypothesis that $\mu_O = \frac{1}{2}(\mu_N + \mu_A)$, we again use a t-test. Using $\bar{x}_O$ to denote the sample mean for Off-type plants, the appropriate statistic is

$$t = \frac{\bar{x}_O - \frac{1}{2}(\bar{x}_N + \bar{x}_A)}{\sqrt{s^2\left\{\dfrac{1}{n_O} + \dfrac{1}{4}\left(\dfrac{1}{n_N} + \dfrac{1}{n_A}\right)\right\}}}.$$

For the given data we find

$$t = \frac{5{\cdot}550 - \frac{1}{2}(6{\cdot}864 + 6{\cdot}462)}{\sqrt{0{\cdot}5536\left\{\dfrac{1}{14} + \dfrac{1}{4}\left(\dfrac{1}{25} + \dfrac{1}{11}\right)\right\}}} = -4{\cdot}64,$$

which we must also compare with the t-distribution on 47 degrees of freedom. The tabulated 0·5% point is approximately 2·69, so the hypothesis will be rejected at the 1% level on a two-tailed test. There is convincing evidence that μ_O is not equal to $\frac{1}{2}(\mu_N + \mu_A)$.

Overall, our conclusion is that the mean rubber contents of Normal and Aberrant plants are very similar, but that they differ from that of Off-type plants.

Calculations

(i) Grand total, $GT = 320{\cdot}39$; number of observations, $n = 50$.

(ii) Correction factor, $CF = (GT)^2/n = 320{\cdot}39^2/50 = 2052{\cdot}995$.

(iii) Totals for types of plant: $T_N = 171{\cdot}61,\ T_O = 77{\cdot}70,\ T_A = 71{\cdot}08$.

Numbers of observations: $n_N = 25,\ n_O = 14,\ n_A = 11$.

(iv) Uncorrected total sum of squares, $USS = 2094{\cdot}5589$.

(v) (Corrected) total sum of squares,
$CSS = USS - CF = 2094{\cdot}5589 - 2052{\cdot}995 = 41{\cdot}5639$.

(vi) Between types sum of squares, $SSB = \dfrac{T_N^2}{n_N} + \dfrac{T_O^2}{n_O} + \dfrac{T_A^2}{n_A} - CF = 15{\cdot}5457$.

(vii) Within types sum of squares $= CSS - SSB = 26{\cdot}0182$.

(viii) Degrees of freedom:

Total $= n - 1 = 49$,

Between types $= 3 - 1 = 2$,

Within types $= 49 - 2 = 47$.

(ix) Mean square = sum of squares / degrees of freedom.

(x) F-ratio $= 7{\cdot}7728/0{\cdot}5536 = 14{\cdot}04$.

Notes

(1) The tests used in the solution require several conditions to be met for their validity. These are that the observations are independent and normally distributed, all with the same variance. In part (a), under the null hypothesis, the values obtained from the 50 plants are a random sample from $N(\mu, \sigma^2)$, where μ is the common value of μ_N, μ_O and μ_A, and σ^2 is the common variance of the observations. Under the alternative, the measurements on the three types of plant form three independent random samples, from normal distributions with different means but with the same variance. (Strictly, the means are not necessarily all different; the alternative hypothesis is just that they are not all the same.)

(2) In the solution to Problem 5B.1 we used the Least Significant Difference method, but have not used it here. This is mainly because, in part (b), there are only two null hypotheses of interest, and because the second of these does not involve a simple comparison of two mean values. The LSD can be used when the aim is to make comparisons between pairs of means. (In fact, its use is rather more complicated when, as here, the sample sizes are unequal.)

* (3) The t-test is useful when testing more complex null hypotheses than those such as $\mu_A = \mu$, where μ is some specified value, or $\mu_N = \mu_A$. A t-test can be generated to test such hypotheses as $a\mu_N + b\mu_O + c\mu_A = 0$, for any given constants a, b and c. In part (b)(ii) we used, in effect, $a = c = -\frac{1}{2}$ and $b = 1$. In these cases the test is conducted by using, as numerator, an estimate of $a\mu_N + b\mu_O + c\mu_A$, viz. $a\bar{x}_N + b\bar{x}_O + c\bar{x}_A$. Now, treating this as a random variable, Y, say, we find that

$$\begin{aligned} \text{Var}(Y) &= \frac{a^2\sigma^2}{n_N} + \frac{b^2\sigma^2}{n_O} + \frac{c^2\sigma^2}{n_A} \\ &= \sigma^2\left(\frac{a^2}{n_N} + \frac{b^2}{n_O} + \frac{c^2}{n_A}\right). \end{aligned}$$

Since, under the hypothesis $a\mu_N + b\mu_O + c\mu_C = 0$, Y is normally distributed and has mean zero,

$$\frac{Y}{\sqrt{\text{Var}(Y)}} \sim N(0,1)$$

when the hypothesis is true. In the usual way, replacing σ^2 by s^2 gives us a statistic with a t-distribution.

5B.3 The quality of electronic components

In a factory manufacturing electronic components there are four machine operators and five machines producing similar items. The quality of the components having been quite variable, an experiment was conducted to determine whether the variability was caused by differences between machines, differences between operators or both. The production was observed for each shift, with each operator using each of the five machines for a whole shift. The quality of the components produced during any shift was assessed by quality control inspectors on a scale of 0 to 100, with 100 corresponding to perfect quality and 0 to useless material. The results were as given in the following table.

Operator	Machine				
	A	B	C	D	E
1	56	92	53	93	68
2	64	83	55	95	62
3	62	80	56	96	62
4	51	78	44	88	69

Carry out an appropriate analysis of variance, and test for differences between operators and between machines.

Solution

In this experiment there are two main sources of variability: differences between operators and differences between machines. The statistical model that is appropriate is, therefore, that for a two-way analysis of variance. The Analysis of Variance table follows, the calculations being presented at the end of the solution.

Source	Degrees of Freedom	Sums of Squares	Mean Squares	F-ratios
Operators	3	129·75	43·25	1·99
Machines	4	4754·30	1188·57	54·75
Residual	12	260·50	21·71	
Total	19	5144·55		

The F-ratio for testing the null hypothesis that there are no differences between operators is 1·99 on 3 and 12 degrees of freedom. This value is compared with upper percentage points of the F-distribution; the 5% point is 3·49, comfortably above the observed value. Hence this null hypothesis cannot be rejected at the 5% level. We cannot, therefore, conclude that the variability in component quality is a result of differences between the operators.

In a similar way the F-ratio for testing the null hypothesis of no differences between the machines is 54·75. This value is highly significant, since it far exceeds 9·63, the 0·1% point of the F-distribution on 4 and 12 degrees of freedom. We conclude that there is overwhelming evidence to reject this null hypothesis. In other words, quality does vary from one machine to another.

Calculations

(i) Grand total, $GT = 1407$; number of observations, $n = 20$.

(ii) Correction factor, $CF = (GT)^2/n = 1407^2/20 = 98\,982{\cdot}45$.

(iii) Totals for operators: $O_1 = 362,\ O_2 = 359,\ O_3 = 356,\ O_4 = 330$.

(iv) Totals for machines: $M_A = 233,\ M_B = 333,\ M_C = 208,\ M_D = 372,\ M_E = 261$.

(v) Uncorrected total sum of squares, $USS = 56^2 + 92^2 + \ldots + 69^2 = 104\,127$.

(vi) (Corrected) total sum of squares,

$$CSS = USS - CF = 104\,127 - 98\,982{\cdot}45 = 5144{\cdot}55.$$

(vii) Number of machines, $n_m = 5$; number of operators, $n_o = 4$.

(viii) Sum of squares between operators, $OSS = \dfrac{O_1^2}{n_m} + \dfrac{O_2^2}{n_m} + \dfrac{O_3^2}{n_m} + \dfrac{O_4^2}{n_m} - CF$

$$= \frac{1}{5}(362^2 + 359^2 + 356^2 + 330^2) - 98\,982{\cdot}45 = 129{\cdot}75.$$

(ix) Sum of squares between machines, $MSS = \dfrac{M_A^2}{n_o} + \dfrac{M_B^2}{n_o} + \dfrac{M_C^2}{n_o} + \dfrac{M_D^2}{n_o} + \dfrac{M_E^2}{n_o} - CF$

$$= \frac{1}{4}(233^2 + 333^2 + 208^2 + 372^2 + 261^2) - 98\,982{\cdot}45 = 4754{\cdot}30.$$

(x) Residual sum of squares,

$$RSS = CSS - OSS - MSS = 5144{\cdot}55 - 129{\cdot}75 - 4754{\cdot}30 = 260{\cdot}50.$$

(xi) Degrees of freedom:

Total $= n - 1 = 19$,

Between operators $= n_o - 1 = 3$,

Between machines $= n_m - 1 = 4$,

Residual $= 19 - 3 - 4 = 12$.

(xii) Mean square = sum of squares / degrees of freedom.

(xiii) F-ratios: $43{\cdot}25/21{\cdot}71 = 1{\cdot}99$; $1188{\cdot}57/21{\cdot}71 = 297{\cdot}14$.

Notes

(1) The statistical model for data such as these is that the observed data value can be expressed as $m + a + b + e$, where m is the overall mean quality level, a is the effect on mean quality caused by using a particular operator, b is the effect on mean quality caused by using a particular machine, and e is a random variable which is normally distributed with mean zero and variance σ^2. Such a model is usually referred to as additive. Note that all data are assumed to have a common variance σ^2. (There are techniques for checking these assumptions, and the analysis can be modified if necessary.)

(2) The data are discrete and are restricted to lie in the range 0-100. Clearly, they cannot be exactly normally distributed. However, data such as those considered here are often approximately normal, and this is usually good enough for practical purposes.

(3) If it is necessary to decide which particular machines differ, the Least Significant Difference (LSD) method can be used. (See Problem 5B.1 for another example of the use of the LSD, and Problem 4B.2 for a note on the theory.) In this problem the LSD, calculated at the 5% significance level, is

$$\text{LSD} = 2{\cdot}18\sqrt{2\times 21{\cdot}71/4} = 7{\cdot}18.$$

The means of the data collected on each machine are, respectively, 58·25, 83·25, 52·00, 93·00 and 65·25. The difference between each pair of means is now compared with the LSD. If the difference between any two means is greater than the LSD, then those means are significantly different at the 5% level. Here all means are significantly different except for those of the pairs (A, C) and (A, E), which have differences of 6·25 and 7·00, respectively.

The formula for the LSD at the 100α% significance level is

$$\text{LSD} = t_{\nu,\alpha/2}\sqrt{2s^2/n_o},$$

where s^2 is the Residual Mean Square, ν being the corresponding number of degrees of freedom, where $t_{\nu,\alpha/2}$ is the appropriate percentage point of the t-distribution on ν degrees of freedom, and n_o is the number of data values used to compute each mean.

* (4) As in Problem 5B.1, we must be aware that when using the LSD we are, in fact, performing six separate t-tests. The 5% significance level should therefore be interpreted with caution. One alternative here would be to use a 1% significance level and obtain better protection against rejecting the null hypothesis when it is true. The only change in the calculations would be to use the t-value for a 1% test, i.e. 3·05. The LSD would then change to 10·05. At this level the means of machines B and D are now also not significantly different. The overall conclusion would now be that machines B and D have a higher quality product than do the other three machines. Of these three, machine A has production quality intermediate between those of C and E.

* (5) Two-way data often exhibit what is termed *interaction*, or non-additivity. In the present example this would imply that the difference in mean quality between two machines depends on which operators are using the machines. To test for non-additivity each operator would need to use each machine for at least two shifts. This would give repeat quality values on each combination of operator and machine, and a model allowing for interaction could be used.

5C Contingency Tables

While one undertakes the analysis of relationships between variables by the techniques of regression and correlation, the topic of contingency tables deals with the analysis of data showing the relationship between attributes. Thus we may wonder to what extent certain social habits (for example, smoking) are related to age, or to sex (or, indeed, to both), or might wish to examine whether cirrhosis of the liver and alcohol consumption are related.

The basic method of data presentation is straightforward. If we have two types of attribute, one with two categories (male, female) and one with three (low, moderate, high) we place each individual into the appropriate one of the six possible categories, thus forming what we call a (2×3) *contingency table* of frequencies. Of course, these numbers and attributes can be generalised, as will be seen in the following problems.

In practice, most contingency tables are many-dimensional, and techniques have been devised recently to analyse such tables. For example, a topic of current interest is whether there is any relationship between presence of heart disease and amount of exercise taken. In an investigation of this question, individuals would also need to be categorised by such factors as age, sex, nationality and type of diet.

5C.1 The value of rainfall forecasts

Forecasts of rainfall are made in three categories, namely 'no rain', 'light rain' and 'heavy rain'. The table below summarises the incidence of each type of forecast and the observed rainfall for a sample of 141 forecasts.

Observed rainfall	Forecast rainfall			
	No rain	Light rain	Heavy rain	Total
No rain	34	24	17	75
Light rain	21	4	3	28
Heavy rain	23	9	6	38
Total	78	37	26	141

(a) Find the 'expected' frequencies for each combination of forecast rainfall and observed rainfall under the hypothesis, H_0, that there is no association between the forecast and the observed rainfall.

(b) Use a test based on the χ^2 distribution to decide whether or not H_0 is plausible.

Solution

(a) We denote the 'expected' frequency of observation for class i of observed rainfall and class j of forecast rainfall by e_{ij}. Then $e_{ij} = n\hat{p}_{i\cdot}\hat{p}_{\cdot j}$, where n is the total number of forecasts, $\hat{p}_{i\cdot}$ is the observed proportion, and hence the estimated probability, of occasions with observed rainfall of type i, and $\hat{p}_{\cdot j}$ is similarly the observed proportion of forecasts of type j. Calculating e_{ij} for the data in the problem gives the following table.

Observed rainfall	Forecast rainfall		
	No rain	Light rain	Heavy rain
No rain	41·5	19·7	13·8
Light rain	15·5	7·3	5·2
Heavy rain	21·0	10·0	7·0

For example, $e_{31} = 141 \times \frac{38}{141} \times \frac{78}{141} = 21{\cdot}0$ (to one decimal place, which is all that is needed here). The other e_{ij}s are obtained similarly.

(b) The form of the test statistic is very similar to that for χ^2 goodness-of-fit tests, as discussed in Chapter 3. We obtain

$$X^2 = \frac{(34 - 41{\cdot}5)^2}{41{\cdot}5} + \frac{(24 - 19{\cdot}7)^2}{19{\cdot}7} + \ldots + \frac{(6 - 7{\cdot}0)^2}{7{\cdot}0}$$

$$= 7{\cdot}84.$$

Under H_0 the test statistic X^2 has, approximately, a χ^2 distribution with $(r-1)(c-1)$ degrees of freedom, where r and c are the numbers of rows and columns in the table. In the present example $r = c = 3$, so that there are 4 degrees of freedom.

As with the χ^2 goodness-of-fit test, H_0 is usually rejected only for *large* values of X^2. The upper 10% and 5% points for χ_4^2 are 7·78 and 9·49, so H_0 would be rejected at the 10%, but not the 5%, significance level. Thus, there is some evidence, although it is not very strong, of association between the forecasts and the observed rainfall.

Notes

(1) In the solution to part (a) we used the expression $e_{ij} = n\hat{p}_{i\cdot}\hat{p}_{\cdot j}$ and showed, as an example, how the formula produces the value 21·0 for e_{31}. As can be seen from the expression for e_{31}, cancellation is possible, and indeed is normally done. A simple general formula for e_{ij} is

$$e_{ij} = \frac{\text{total for row } i \times \text{total for column } j}{\text{grand total}},$$

so that, for example,

$$e_{31} = \frac{38 \times 78}{141} = 21{\cdot}0\ .$$

(2) The conclusion to part (b) was that the data contained some slight evidence against H_0, i.e. in favour of an association between the forecast and the actual weather. In practice, the statistician would now go on, in consultation with specialists in the topic, to consider what form any association between these categories might take. In the present case, it is interesting to note that, if anything, the association is the reverse of the one to be expected. For example, when *heavy* rain is forecast the observed frequency of *no* rain is greater than its expected frequency.

5C.2 A survey of exercise habits

In a survey, 1000 individuals were asked how often they took exercise in the form of cycling. There were three possible responses, namely 'never', 'occasionally' and 'frequently'. A similar question was asked with respect to walking, with the same three possible responses.

Of the 1000 individuals, 75 said that they frequently walked and cycled, 10 never walked but cycled frequently, 35 never cycled but walked frequently, and 100 never walked and never cycled. In all, 200 individuals never cycled and 125 never walked, whereas 700 cycled occasionally and 150 walked frequently.

(a) Arrange the data in the form of a (3×3) contingency table, and state how many of the 1000 individuals both cycle and walk occasionally.

(b) Apply a χ^2 test to the (3×3) table, stating clearly the hypothesis which it is designed to test.

Solution

(a) The (3×3) table is given below. The entries marked by an asterisk are given in the problem, and the remaining entries follow by subtraction.

		Cycling			
		Never	Occasionally	Frequently	Total
Walking	Never	100*	15	10*	125*
	Occasionally	65	645	15	725
	Frequently	35*	40	75*	150*
	Total	200*	700*	100	1000*

Hence the number of individuals who both cycle and walk occasionally is 645.

(b) The usual χ^2 statistic is

$$\sum_{i=1}^{3}\sum_{j=1}^{3}\frac{(f_{ij}-e_{ij})^2}{e_{ij}},$$

where the f_{ij}s are the entries in the table of observed frequencies constructed in part (a) and the e_{ij}s are the corresponding expected frequencies under the null hypothesis. The statistic is designed to test the null hypothesis that the variables defining the margins of the table (in this case the amounts of cycling and walking) are independent, against a general alternative. The e_{ij}s are calculated as in Problem 5C.1, so that, for example,

$$e_{23} = n\hat{p}_{2\cdot}\hat{p}_{\cdot 3} = 1000\times\frac{725}{1000}\times\frac{100}{1000} = 72{\cdot}5.$$

A complete table of e_{ij}s is given below.

		Cycling			
		Never	Occasionally	Frequently	Total
Walking	Never	25·0	87·5	12·5	125
	Occasionally	145·0	507·5	72·5	725
	Frequently	30·0	105·0	15·0	150
	Total	200	700	100	1000

The test statistic is

$$X^2 = \frac{(100-25{\cdot}0)^2}{25{\cdot}0}+\frac{(15-87{\cdot}5)^2}{87{\cdot}5}+\frac{(10-12{\cdot}5)^2}{12{\cdot}5}+\ldots+\frac{(75-15{\cdot}0)^2}{15{\cdot}0} = 693{\cdot}64.$$

Under the null hypothesis of independence, the test statistic X^2 has a χ^2 distribution with $(r-1)(c-1) = 4$ degrees of freedom. The value of X^2, 693·64, greatly exceeds any of the usual percentage points for χ_4^2. We therefore conclude that amounts of walking and cycling are not independent.

Notes

(1) The χ^2 test for contingency tables, as with the χ^2 goodness-of-fit test, can detect any sort of alternative to the null hypothesis. Thus the χ^2 test will have some chance of detecting any dependence between the two variables, no matter what form this might take. However, in some circumstances only one particular type of dependence is of major interest as an alternative to independence. This is true, in particular, when the categories of the two variables are ordered, and it is anticipated that they will increase or decrease together. This would happen in the present example if it were thought that the alternative to independence is that amounts of exercise in the two activities increase together. We would probably not, for example, consider an alternative where those who occasionally cycled rarely walked, but those in the other two categories, who 'never cycled' or 'frequently cycled', walked more than average; such an

alternative would however be detected by a χ^2 test.

If the range of plausible alternatives is restricted, then there may be other tests which are better able to detect them (are more powerful) than the χ^2 test. This increased power is obtained by losing the ability to detect alternatives outside the limited range of interest.

In the present example the evidence of dependence is so strong that there is no need to consider tests other than χ^2. However, in tables with ordered categories where the χ^2 test just fails to reject independence, a more specific test based on restricted alternatives may provide conclusive evidence against independence. Note, however, that it is not acceptable to 'snoop' through the data, trying to discover a plausible alternative hypothesis, and then choose the test most likely to reject the null hypothesis in favour of that alternative. In principle, one ought to decide on the test to be used before collecting the data. (A similar point is made in the discussion of Problem 5A.2, in relation to choice of a one-tailed or two-tailed test in a different context.)

(2) It is really not necessary to calculate the value of the χ^2 statistic precisely in this example, unless an exact significance level (or p-value) is required. The first term in the expression for the statistic is 225, which on its own greatly exceeds any of the usual χ^2 percentage points.

* 5C.3 Sex ratios of insects

(a) In a contingency table with r rows and 2 columns the observations in row i are n_{i1} and n_{i2}, with row totals $n_{i.} = n_{i1} + n_{i2}$, $i = 1, 2, \ldots, r$; the column totals are $n_{.1}$ and $n_{.2}$, with $n_{.1} + n_{.2} = n$, the total number of observations. Show that the χ^2 statistic for testing independence of the variables defining the rows and columns of the table can be written as

$$X^2 = \frac{n^2}{n_{.1}n_{.2}} \sum_{i=1}^{r} \frac{(n_{i1} - n_{i.}n_{.1}/n)^2}{n_{i.}}.$$

Hence, or otherwise, show that it can be expressed as

$$X^2 = \frac{n^2}{n_{.1}n_{.2}} \left\{ \sum_{i=1}^{r} \frac{n_{i1}^2}{n_{i.}} - \frac{n_{.1}^2}{n} \right\}.$$

(b) Samples of a certain species of insect were collected from two different locations, and the number of females was observed. There were 44 females in the 100 insects collected at the first location, and 86 females out of 200 insects at the second location. Test whether the proportions of females differ between the two locations.

(c) A further sample of 200 insects is taken at a third location and found to contain 110 females. Consider the data from all three locations simultaneously, and test whether there are differences in the proportions of females between the three locations.

Solution

(a) The χ^2 statistic is given in general by

$$X^2 = \sum_{i=1}^{r} \sum_{j=1}^{2} \frac{(f_{ij} - e_{ij})^2}{e_{ij}} = \sum_{i=1}^{r} S_i,$$

say, and in the present case $f_{ij} = n_{ij}$ and $e_{ij} = n_{i.}n_{.j}/n$, $i = 1, 2, \ldots, r$, $j = 1, 2$. Now

$$\begin{aligned} S_i &= \sum_{j=1}^{2} \frac{(f_{ij} - e_{ij})^2}{e_{ij}} \\ &= \frac{(n_{i1} - n_{i.}n_{.1}/n)^2}{n_{i.}n_{.1}/n} + \frac{(n_{i2} - n_{i.}n_{.2}/n)^2}{n_{i.}n_{.2}/n}. \end{aligned}$$

But

$$n_{i2} - n_{i\cdot}n_{\cdot 2}/n = n_{i\cdot} - n_{i1} - n_{i\cdot}(n - n_{\cdot 1})/n$$
$$= n_{i\cdot} - n_{i1} - n_{i\cdot} + n_{i\cdot}n_{\cdot 1}/n$$
$$= -(n_{i1} - n_{i\cdot}n_{\cdot 1}/n).$$

Hence

$$S_i = \frac{(n_{i1} - n_{i\cdot}n_{\cdot 1}/n)^2}{n_{i\cdot}/n}\left[\frac{1}{n_{\cdot 1}} + \frac{1}{n_{\cdot 2}}\right]$$
$$= \frac{n^2(n_{i1} - n_{i\cdot}n_{\cdot 1}/n)^2}{n_{i\cdot}n_{\cdot 1}n_{\cdot 2}},$$

so that

$$X^2 = \frac{n^2}{n_{\cdot 1}n_{\cdot 2}}\sum_{i=1}^{r}\frac{(n_{i1} - n_{i\cdot}n_{\cdot 1}/n)^2}{n_{i\cdot}}, \text{ as required.}$$

We now wish to obtain the second expression for X^2. Expanding the numerator, we find that

$$\sum_{i=1}^{r}\frac{(n_{i1} - n_{i\cdot}n_{\cdot 1}/n)^2}{n_{i\cdot}} = \sum_{i=1}^{r}\frac{n_{i1}^2}{n_{i\cdot}} - \frac{2n_{\cdot 1}}{n}\sum_{i=1}^{r}n_{i1} + \frac{n_{\cdot 1}^2}{n^2}\sum_{i=1}^{r}n_{i\cdot}$$
$$= \sum_{i=1}^{r}\frac{n_{i1}^2}{n_{i\cdot}} - \frac{2n_{\cdot 1}^2}{n} + \frac{n_{\cdot 1}^2}{n}, \text{ since } \sum_{i=1}^{r}n_{i\cdot} = n \text{ and } \sum_{i=1}^{r}n_{i1} = n_{\cdot 1},$$
$$= \sum_{i=1}^{r}\frac{n_{i1}^2}{n_{i\cdot}} - \frac{n_{\cdot 1}^2}{n}.$$

We thus obtain

$$X^2 = \frac{n^2}{n_{\cdot 1}n_{\cdot 2}}\left\{\sum_{i=1}^{r}\frac{n_{i1}^2}{n_{i\cdot}} - \frac{n_{\cdot 1}^2}{n}\right\},$$

the required expression.

(b) This can be viewed either as a test of equality of two binomial parameters or as a (2×2) contingency table. These two viewpoints lead to apparently different, but in fact equivalent, test statistics, and we will give both methods, for completeness.

Method 1

Let p_1 and p_2 be the proportions of female insects in the populations at the two locations. We then wish to test H_0: $p_1 = p_2$ against H_1: $p_1 \neq p_2$. If $\hat{p}_1$ and $\hat{p}_2$ are the sample proportions of females, then using the normal approximation to the binomial distribution we have, approximately,

$$\hat{p}_i \sim N(p_i, p_i q_i/n_i), \quad i = 1, 2,$$

where $q_i = 1 - p_i$ and n_i is the number of observations (or trials) in sample i. Assuming that the two samples are independent, it follows that, approximately,

$$\hat{p}_1 - \hat{p}_2 \sim N\left(p_1 - p_2, \frac{p_1 q_1}{n_1} + \frac{p_2 q_2}{n_2}\right),$$
$$\text{i.e. } \hat{p}_1 - \hat{p}_2 \sim N\left(0, \frac{pq}{n_1} + \frac{pq}{n_2}\right)$$

under H_0, where p is the common value of p_1 and p_2 under this hypothesis, and $q = 1 - p$.

The value of p is of course unknown, but the obvious estimate is $\hat{p}$, the proportion of females in the two samples together. We have then that, approximately,

$$Z = \frac{\hat{p}_1 - \hat{p}_2}{\left[\hat{p}\hat{q}\left(\frac{1}{n_1} + \frac{1}{n_2}\right)\right]^{\frac{1}{2}}} \sim N(0,1),$$

when H_0 is true, and Z can be used as our test statistic.

For the given data, $\hat{p}_1 = 0{\cdot}440$, $\hat{p}_2 = 0{\cdot}430$, $\hat{p} = 0{\cdot}433$, $n_1 = 100$, $n_2 = 200$, so

$$Z = \frac{0{\cdot}440 - 0{\cdot}430}{\left[0{\cdot}433 \times 0{\cdot}567 \times \left(\frac{1}{100} + \frac{1}{200}\right)\right]^{\frac{1}{2}}} = \frac{0{\cdot}010}{[0{\cdot}003\,68]^{\frac{1}{2}}}$$

$$= 0{\cdot}165.$$

Since this value is to be compared with percentage points of $N(0, 1)$ on a two-tailed test, for example, $-1{\cdot}96$ and $+1{\cdot}96$ for a test of size 5%, we conclude that there is no evidence of any difference between p_1 and p_2. This might have been expected, given the obvious closeness of the two sample proportions $\hat{p}_1$ and $\hat{p}_2$.

Method 2

Expressing the data as a contingency table we obtain the following.

		Location 1	Location 2	Total
Sex	Female	44	86	130
	Male	56	114	170
	Total	100	200	300

The $\{e_{ij}\}$ needed in the χ^2 test statistic X^2 are given by

$$e_{11} = \frac{130 \times 100}{300} = 43\tfrac{1}{3}$$

$$\text{and } e_{12} = \frac{130 \times 200}{300} = 86\tfrac{2}{3},$$

and similarly $e_{21} = 56\frac{2}{3}$ and $e_{22} = 113\frac{1}{3}$. (Since the numbers in the 'Total' row and column are given, the last three of these could have been calculated by subtraction; for example, $e_{12} = 130 - e_{11}$.) The χ^2 statistic X^2 is therefore

$$X^2 = \frac{(44 - 43\frac{1}{3})^2}{43\frac{1}{3}} + \frac{(86 - 86\frac{2}{3})^2}{86\frac{2}{3}} + \frac{(56 - 56\frac{2}{3})^2}{56\frac{2}{3}} + \frac{(114 - 113\frac{1}{3})^2}{113\frac{1}{3}}$$

$$= \left(\frac{2}{3}\right)^2 \left[\frac{1}{43\frac{1}{3}} + \frac{1}{86\frac{2}{3}} + \frac{1}{56\frac{2}{3}} + \frac{1}{113\frac{1}{3}}\right] = 0{\cdot}027.$$

This is, in fact, simply the square of Z calculated above. Since $Z \sim N(0, 1)$ approximately, it follows that $Z^2 \sim \chi_1^2$ approximately, so we have here separate confirmation that the statistic X^2 for the χ^2 test does indeed have, approximately, a χ^2 distribution. It is clear, as with Method 1 above, that the statistic provides no evidence to contradict the hypothesis of independence between location and proportion of females.

(c) We now have a (2×3) contingency table, as follows.

		Location 1	2	3	Total
Sex	Female	44	86	110	240
	Male	56	114	90	260
	Total	100	200	200	500

The expected frequency e_{11} of females in location 1, assuming independence between location and proportion of females, is given by

$$e_{11} = \frac{240\times100}{500} = 48{\cdot}0,$$

and the remaining expected numbers can be calculated similarly (with some alternatively obtained by subtraction) to give the following table of expected frequencies.

		Location 1	2	3	Total
Sex	Female	48	96	96	240
	Male	52	104	104	260
	Total	100	200	200	500

The χ^2 test statistic X^2 then takes the value

$$X^2 = \frac{(44-48)^2}{48} + \frac{(86-96)^2}{96} + \ldots + \frac{(90-104)^2}{104} = 6{\cdot}57.$$

The number of degrees of freedom is $(r-1)\times(c-1) = 1\times2$, and the 5% and $2\frac{1}{2}$% points of χ_2^2 are 5·99 and 7·38 respectively. There is thus evidence at the 5% level, but not at the $2\frac{1}{2}$% level, that the proportions of females in the three locations are different.

Notes

(1) Algebraic manipulations similar to those in part (a) of the solution show the equivalence of Z^2 and X^2, the statistic for the χ^2 test. There is yet another expression for X^2 which is available for (2×2) tables and is simpler to calculate than the more general expressions. This is

$$X^2 = \frac{n(n_{11}n_{22} - n_{12}n_{21})^2}{n_{1\cdot}n_{2\cdot}n_{\cdot1}n_{\cdot2}},$$

where $n_{1\cdot}$, $n_{2\cdot}$, $n_{\cdot1}$ and $n_{\cdot2}$ are the marginal totals of the table, and n is the grand total. It is usually quoted in the alternative notation

$$X^2 = \frac{n(ad-bc)^2}{(a+b)(c+d)(a+c)(b+d)},$$

where a, b, c and d are the four observed frequencies n_{11}, n_{12}, n_{21} and n_{22} respectively. Substituting $e_{ij} = n_{i\cdot}n_{\cdot j}/n$ into the general expression, and some algebraic manipulation, quickly leads to the alternative expression. In the present example the alternative expression gives

$$\frac{300\ (44\times114 - 86\times56)^2}{130\times170\times100\times200} = 0{\cdot}027, \text{ as before.}$$

(2) The distribution of the test statistic calculated for contingency tables is only approximately the χ^2 distribution. As with goodness-of-fit tests, the main restriction needed to ensure a good

approximation is that none of the e_{ij}s is too small, where 'small' takes on roughly the same meaning as for goodness-of-fit tests. (See Note 3 to Problem 3B.2.)

For (2×2) tables the approximation can be improved by replacing $(f_{ij} - e_{ij})^2$ in the test statistic by $(|f_{ij} - e_{ij}| - \frac{1}{2})^2$. This modification is termed *Yates' correction*, and is a special case of a continuity correction. Continuity corrections are appropriate whenever a discrete distribution is approximated by a continuous distribution, and the exact distribution of the χ^2 test statistic is discrete, whereas of course the χ^2 distribution is continuous.

Continuity corrections are, in theory, also desirable for larger tables, but they are less likely to make much difference to the value of the test statistic and they do not take a simple form, so are generally ignored. Even for (2×2) tables, Yates' correction will often make very little difference to the result. Also, the correction in fact tends to 'over-correct'. Without the correction there is a tendency for H_0 to be rejected slightly too often, whereas the opposite is the case (the test being said to become *conservative*) when the correction is used.

For (2×2) tables, if the marginal totals are not very large, it is possible to calculate the exact distribution of the test statistic, rather than relying on the χ^2 approximation. The exact distribution is related to the hypergeometric distribution, which is the distribution of any one of the individual cell values. (See Problem 2A.7 for a discussion of this distribution.) Calculating the exact distribution becomes very tedious when the marginal totals are at all sizeable, which is in any case when the χ^2 approximation will be a good one.

(3) Given the conclusions in parts (b) and (c) it seems likely that locations 1 and 2 both differ from location 3, but not from one another. If this alternative were of interest, rather than the very general alternative of there being *some* difference between the three locations, a more powerful test is available by combining the first two columns of the contingency table. If this is done we reach the table below (with expected frequencies in parentheses).

		Location				
		1 and 2		3		Total
Sex	Female	130	(144·0)	110	(96·0)	240
	Male	170	(156·0)	90	(104·0)	260
	Total	300		200		500

The test statistic now takes the value

$$14^2\left(\frac{1}{144} + \frac{1}{96} + \frac{1}{156} + \frac{1}{104}\right) = 6{\cdot}54,$$

or, incorporating Yates' correction,

$$(13{\cdot}5)^2\left(\frac{1}{144} + \frac{1}{96} + \frac{1}{156} + \frac{1}{104}\right) = 6{\cdot}08.$$

These values lie between the $2\frac{1}{2}\%$ (5·02) and 1% (6·63) critical values of χ_1^2, so that the evidence is moderate, but still not conclusive, that there are differences between the locations. The present test is, however, somewhat more powerful than the general one presented earlier.

(4) In the solution to part (b), Method 1, we followed the standard, and intuitively sensible, practice, of replacing p_1 and p_2 by a common estimate $\hat{p}$ rather than by separate estimates $\hat{p}_1$ and $\hat{p}_2$ when the null hypothesis is $H_0: p_1 = p_2$. However, some authors use separate estimators, and such use cannot be ruled out as completely unacceptable.

5D Time Series

One of the types of statistical information which come most readily to the public eye is that contained in regular government publications. Thus, for example, most governments publish monthly a great number of economic and social statistics; unemployment figures and balance of trade figures are perhaps the most prominent, but there are many more. These statistics, giving the successive values taken by some variable at more or less regular intervals of time, are known as *time series*, and while the discussion above has concentrated on the prominent governmental area, applications of time series analysis occur much more widely. For example, meteorological data have been kept regularly over many years, so that the analysis of time series in climatology is a standard technique. Another example is found in the area of dendrochronology, the dating of wooden objects through the analysis of the widths of successive tree-rings; these tree-rings are formed by the annual growth of bark on a tree, and the width reflects (amongst other things) rainfall, so that matching of rings from different trees through the methods of time series is a practical method.

Time series analysis is, however, rather a difficult subject, and seldom encountered in examinations at the level we have been considering in this book. We end with two representative, but rather different, time series problems, as an introduction to this interesting topic.

5D.1 The electricity consumption of a Canadian household

(a) In the analysis of a time series, explain what are meant by the terms *trend, seasonal variation* and *random variation*.

(b) For a period of 19 weeks, from mid-November until the end of March, the weekly electricity consumption of a Canadian household was recorded, and the results are given below. Plot the data on a graph.

Week	Units of electricity	Week	Units of electricity
1	65	11	74
2	73	12	75
3	72	13	73
4	68	14	67
5	73	15	67
6	86	16	62
7	79	17	60
8	87	18	63
9	82	19	62
10	82		

(c) Calculate an appropriate moving average in order to smooth the series, and plot its values on the graph.

(d) Plot, separately, the remainder left when the moving average is subtracted from the observed series, and discuss how the graphs in (c) and (d) are related to the ideas of trend, seasonal variation and random variation.

(e) Discuss, with reasons, what you think might happen to electricity consumption during the next 20 weeks after the end of the series given in (b).

Solution

(a) A traditional approach to the analysis of time series is to imagine that an observed series is the sum (or perhaps the product) of three separate parts, as below.

(i) A slowly-varying, 'smooth' part of the series which represents the long-term *trend*.

(ii) A periodic component which repeats itself regularly. This *seasonal variation* usually has a period of one year and is caused by the different nature of environmental or human behaviour at different times of year. The term *seasonal variation* is, however, sometimes used to represent other periodic behaviour with a fixed period, for example diurnal variation.

(iii) *Random variation* is essentially what remains after trend and seasonal variation have been removed from a series. It is not necessarily random in the strict sense that successive values are independent. Typically, indeed, they are not independent, and this random component of a series can include, for example, behaviour which is almost periodic in nature.

(b) The required graph is shown in Figure 5.5.

(c) There are several moving averages which could be used. If we denote the series by x_t, $t = 1, 2, \ldots, 19$, and denote the moving average by y_t, then we could consider using the following:

(i) a simple three-point moving average

$$y_t = \tfrac{1}{3}(x_{t-1} + x_t + x_{t+1}), \quad t = 2, 3, \ldots, 18;$$

(ii) a simple five-point moving average

$$y_t = \tfrac{1}{5}(x_{t-2} + x_{t-1} + x_t + x_{t+1} + x_{t+2}), \quad t = 3, 4, \ldots, 17;$$

(iii) a three-point moving average with linearly decreasing weights, i.e. weights which decrease linearly on either side of t

$$y_t = \tfrac{1}{4}(x_{t-1} + 2x_t + x_{t+1}), \quad t = 2, 3, \ldots, 18;$$

(iv) a five-point moving average with linearly decreasing weights

$$y_t = \tfrac{1}{9}(x_{t-2} + 2x_{t-1} + 3x_t + 2x_{t+1} + x_{t+2}), \quad t = 3, 4, \ldots, 17.$$

These are by no means the only possibilities, but any of them would be adequate for the present problem. A moving average is generally based on an odd number of time points, since using an even number will result in an expression for y_t for a fractional value of t. This would be inconvenient when one is trying to compare x_t and y_t.

Only *one* of the above moving averages is required here, but for illustration we calculate y_t using two of them, (ii) and (iii). These will give, respectively, the most drastic and least drastic smoothing of the series, among the four listed.

Week	y_t (ii)	y_t (iii)	Week	y_t (ii)	y_t (iii)
2		70·75	11	77·2	76·25
3	70·2	71·25	12	74·2	74·25
4	74·4	70·25	13	71·2	72·0
5	75·6	75·0	14	68·8	68·5
6	78·6	81·0	15	65·8	65·75
7	81·4	82·75	16	63·8	62·75
8	83·2	83·75	17	62·8	61·25
9	80·8	83·25	18		62·0
10	80·0	80·0			

Both these moving averages are plotted on the graph in Figure 5.5; it can be seen that they give similar, though certainly not identical, pictures.

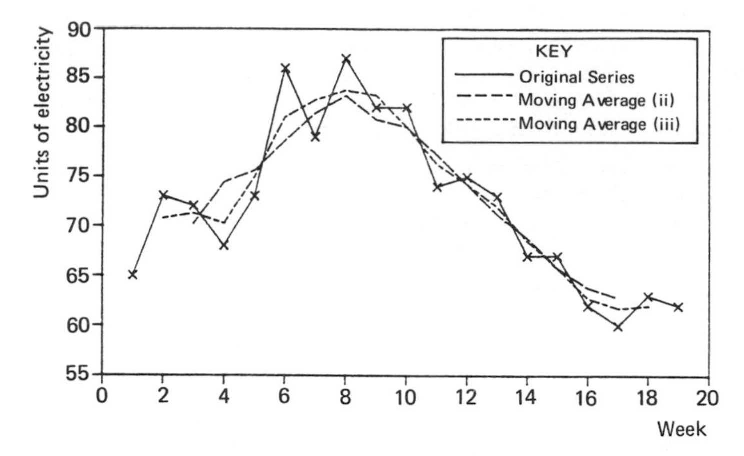

Figure 5.5 Electricity consumption data from Problem 5D.1

(d) For the two chosen moving averages, the remainder series $e_t = x_t - y_t$ are as given in the following table.

Week	e_t (ii)	e_t (iii)	Week	e_t (ii)	e_t (iii)
2		2·25	11	−3·2	−2·25
3	1·8	0·75	12	0·8	0·75
4	−6·4	−2·25	13	1·8	1·0
5	−2·6	−2·0	14	−1·8	−1·5
6	7·4	5·0	15	1·2	1·25
7	−2·4	−3·75	16	−1·8	−0·75
8	3·8	3·25	17	−2·8	−1·25
9	1·2	−1·25	18		1·0
10	2·0	2·0			

The plots shown in Figure 5.6 of these two series of $\{e_t\}$ are very similar, although the second, which smooths less drastically than the first, follows the original series more closely and so tends to have smaller values of $|e_t|$.

Moving averages may be used to estimate the trend in a series, since they give a smooth version of the series and trend is defined as the 'smooth part' of any series. In this example the 'trend' is in reality part of a seasonal cycle with maximum electricity consumption occurring in December and January. However to identify a separate seasonal part in a series, it is necessary to have two or more complete cycles. In the present case where only part of a cycle is available, any seasonal variation will appear as part of the trend.

The remainder series, e_t, therefore estimates the extent of random variation. In the present example, there is some evidence of a pattern among successive values of e_t in that an increase in e_t at point t has a tendency to be followed by a decrease, and vice versa. However, with such a short series the evidence is not conclusive, and it is possible that e_t is a series of independent, identically distributed random variables (i.e. e_t is a random series in the strict sense of the word).

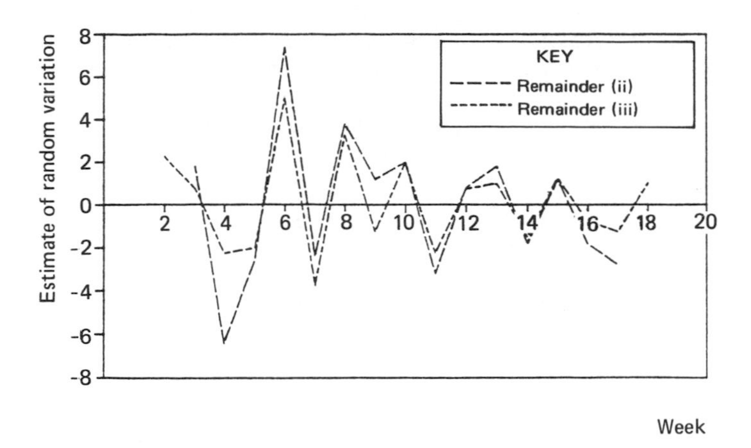

Figure 5.6 Plots of remainder series for Problem 5D.1

(e) It is always dangerous to extrapolate (forecast) a series without additional information, apart from the observed values of the series, of what might happen in the future. In the present example, the graph is ambiguous; the consumption could be part of the way through a long decline, or the decline may be showing signs of levelling off. From the context of the problem, it is likely that electricity consumption will continue to fall for another month or two as heating and lighting requirements decline, whereas in late summer (towards the end of the 20-week forecast period) the consumption will probably begin to rise again. There could also be a secondary peak in the middle of summer, because of additional refrigeration requirements.

Note

The choice of a moving average to use is rather subjective, and the main consideration is the amount of smoothing which is required. Typically, a moving average will become smoother if

(a) the number of terms used in the average is increased, and

(b) the weights are all equal, not decreasing either side of t.

Thus, in the solution, moving average (ii) gives the smoothest result of the four averages mentioned, and (iii) gives the least smooth.

5D.2 Turning points in a time series

A time series consists of observations $x_1, x_2, \ldots, x_n$ taken on a random variable at equally-spaced points of time. If, of three consecutive observations x_{t-1}, x_t and x_{t+1}, x_t is the smallest or largest of the three, then there is said to be a *turning-point* in the series at time t. Let S be the total number of turning-points in the series of n observations.

(a) For the case $n = 4$, suppose that x_1, x_2, x_3 and x_4 take the values 1, 2, 3 and 4 in some order. Write down all the permutations of these four numbers, and for each permutation write down the value which S would take if x_1 were the first number, x_2 the second, and so on. If all permutations are equally likely, what is the probability function of S?

(b) Now suppose that $x_1, x_2, \ldots, x_n$ are independent observations, all from the same continuous probability distribution. Show that the expected value of S is $\frac{2}{3}(n-2)$.

Solution

(a) There are 24 permutations of the numbers 1, 2, 3 and 4. These are set out in the table below, together with the associated values taken by S, the number of turning-points.

Permutation	S	Permutation	S
1234	0	3124	1
1243	1	3142	2
1324	2	3214	1
1342	1	3241	2
1423	2	3412	2
1432	1	3421	1
2134	1	4123	1
2143	2	4132	2
2314	2	4213	1
2341	1	4231	2
2413	2	4312	1
2431	1	4321	0

Amongst the 24 permutations, the frequency of '$S=0$' is 2, that of '$S=1$' is 12 and that of '$S=2$' is 10. Since the permutations are given to be equally likely, each has probability $\frac{1}{24}$, and the probability distribution of S is thus as in the table below.

s	0	1	2
$\Pr(S=s)$	$\frac{1}{12}$	$\frac{1}{2}$	$\frac{5}{12}$

(b) The relative positions of the xs when arranged in order of magnitude determine the value of S. Given these positions, the actual values taken by the xs are irrelevant to S. Note also that, because the distribution of the xs is continuous, ties (i.e. instances of equality between two or more xs) can be assumed to have probability zero.

Consider now three consecutive observations x_{t-1}, x_t and x_{t+1}. These can be permuted in 3!, or 6, ways. For four of these, viz. $\{x_t < x_{t-1} < x_{t+1}\}$, $\{x_t < x_{t+1} < x_{t-1}\}$, $\{x_{t+1} < x_{t-1} < x_t\}$ and $\{x_{t-1} < x_{t+1} < x_t\}$, there is a turning-point at time t, and for the other two there is not. Because the observations are independent and identically distributed, each of the six permutations has the same probability, so we obtain

$$\Pr(\text{turning-point at } t) = \tfrac{4}{6} = \tfrac{2}{3}.$$

Now $S = S_2 + S_3 + \ldots + S_{n-1}$, where, for $t = 2, 3, \ldots, n-1$, S_t is defined as 1 if there is a turning-point at time t, and as 0 otherwise. Then

$$E(S) = \sum_{t=2}^{n-1} E(S_t),$$

and since

$$E(S_t) = 0 \times \Pr(S_t = 0) + 1 \times \Pr(S_t = 1) = \tfrac{2}{3}, \quad t = 2, 3, \ldots, n-1,$$

we obtain

$$E(S) = \tfrac{2}{3}(n-2).$$

Notes

(1) This problem is concerned with finding a probability distribution and an expectation, and could therefore have been included in Section 1B. However, it is presented here in the context of the analysis of a series of observations in time, i.e. a time-series, and so is relevant to the present section.

(2) The result $E(S) = \sum E(S_t)$, used in solving part (b), does not require $S_2, S_3, \ldots, S_{n-1}$ to be independently distributed and, indeed, they are not. This lack of independence means that we could not use a similar argument to obtain $\text{Var}(S)$.

(3) Turning-points provide one way of detecting whether or not a time series is 'purely random', in the sense that all observations are independent and identically distributed. The alternatives to randomness are often a steadily increasing or decreasing trend, cyclic behaviour or some other *structure* underlying the series.

For most structured time series, the number of turning-points will be less than would be expected for a purely random series. Given the distribution of S for a purely random series (found here for $n = 4$) a test of the hypothesis of randomness can therefore be carried out which will reject the hypothesis for particularly small values of S.

Such a test could also be used in connection with a regression analysis. One of the assumptions in regression is that the 'random errors' are independent and identically distributed. One can test this by examining the residuals, i.e., the discrepancies between the observed values and the predicted values. (See Problem 5A.1 for an example.) Plotting these residuals against values of time or some other potential explanatory variable can be used to test the 'randomness' of the errors.

(4) The test based upon turning-points, like several other tests for randomness, is capable of detecting many different forms of structure in a time series. However, if one particular type of structure, such as a linear trend, is of particular interest, other, more specialised, tests will usually have greater power.

Index

Items are referred to by problem number or by page number as appropriate. Entries in **bold** type are problem numbers, those in ordinary type are page numbers. Problem numbers are used to index titles, and topics of central importance. For certain topics, only some of the more important occurrences are indexed.